A LIFE IN SCIENCE

C.N.R. RAO

A LIFE IN SCIENCE

PENGUIN
VIKING
An imprint of Penguin Random House

VIKING

USA | Canada | UK | Ireland | Australia
New Zealand | India | South Africa | China | Singapore

Viking is part of the Penguin Random House group of companies
whose addresses can be found at global.penguinrandomhouse.com

Published by Penguin Random House India Pvt. Ltd
4th Floor, Capital Tower 1, MG Road,
Gurugram 122 002, Haryana, India

First published in Viking by Penguin Random House India 2016

10 9 8 7 6 5 4 3 2

ISBN 9780670089093

Typeset in Adobe Garamond Pro by Manipal Digital Systems, Manipal
Printed at Replika Press Pvt. Ltd, India

www.penguin.co.in

This is a legitimate digitally printed version of the book and therefore might not
have certain extra finishing on the cover.

*This book is warmly dedicated to aspiring young scientists
and to all those who want to make friends with science*

CONTENTS

PROLOGUE

Reaching the age of eighty is considered somewhat special in India. One is supposed to have seen a thousand moons. I am now well past eighty and have certainly had an interesting academic life. I have been urged by some friends and publishers to write a book describing how and why I developed an abiding interest in science and other academic pursuits, and have been able to contribute to scientific research for over six decades. A book somewhat autobiographical in nature, dealing with the agony and ecstasy in scientific pursuit was considered to be timely. Such a book was to be especially addressed to young people, a majority of whom may be looking for venues that satisfy their intellectual aspirations and also provide fruitful occupation. This was considered important in view of the confusing educational scenario and the absence of obvious opportunities for creative, young people. Youngsters, as well as teachers, need to be motivated to tread paths leading to creative exploration in science and other areas. This is necessary for the future of the country, and the well-being of our society and humanity as a whole. I decided to take up the task of writing this book not necessarily because it may be timely and interesting, but more because I may not be able to write it after a few years. I have taken some liberty in writing about various matters freely, assuming the advantage of age.

Hearing great scientists or reading about them has motivated many to take up science. I myself have been inspired by the lives and work of great scientists, some of whom carried out monumental work in spite of the many impediments in their way. The book starts with autobiographical accounts by six well-known scientists as to why and how they took up science (Chapter 1). In Chapter 2, the nature of science and its attributes are presented briefly. My journey in science is covered in Chapters 3 to 6, starting from my early years to the present day. I have found it convenient to write the chapters dealing with the different phases of my life in science, which interestingly coincided with the different periods of my life. Chapter 3 deals with the different stages of my education—from middle school to the postdoctoral level. An account of the tribulations of the first twenty-five years of my career is presented in Chapter 4. Chapter 5 gives an account of the busy period between 1984 and 1994, when I had to work in different spheres simultaneously and yet carry out research with my students. This period demanded judicious management of time. After I gave up much of the formal administrative responsibilities, I have enjoyed science in a more relaxed fashion and have contributed to institution-building to some extent (Chapter 6). It is noteworthy that my research performance, which showed a significant improvement after my sixtieth year, became even better after my seventieth year—the impact index and number of citations reached levels beyond my expectations.

I write about my heroes in science in Chapter 7. In the Epilogue, I present my personal reflections on science, education and other aspects, expressing my hopes and fears. Throughout the text, I have written about several events in my life for the sake of completeness, and for readers to view the various events within a proper framework and to mark the passage of time.

Appendix 1 is a letter to a young friend, which tries to answer questions and doubts raised by most students related to educational opportunities and their choice of study. The book ends with a short account of how many great scientists gave their all to science and to young co-workers.

I trust that readers will find the book interesting and useful. I will feel more than rewarded if this is so.

I express my sincere thanks to Ms Shashi Karthikeyan for her invaluable assistance in preparing the manuscript. My thanks are due to my wife, Indu, for reading the manuscript and for her helpful suggestions. My special thanks to Jatinder Kaur, who was of great support in formatting the manuscript.

ON AUTOBIOGRAPHICAL WRITING

(From www.nobelprize.org)

Here I sit in order to write, at the age of sixty-seven, something like my own obituary. I am doing this not merely because Dr Schilpp has persuaded me to do it, but because I do, in fact, believe that it is a good thing to show those who are striving alongside of us how our own striving and searching appears in retrospect. After some reflection, I felt how imperfect any such attempt is bound to be. For however brief and limited one's working life may be, and however predominant may be the way of error, the exposition of that which is worthy of communication does nonetheless not come easy. Today's person of sixty-seven is by no means the same as was the one of fifty, of thirty or of twenty. Every reminiscence is coloured by one's present state, hence by a deceptive point of view. This consideration could easily deter one. Nevertheless, much can be gathered out of one's own experience that is not open to another consciousness—**Albert Einstein**

1. WHY BE A SCIENTIST?

It is motivating and exciting to learn about scientists, and their science, in their own words. In this chapter, short autobiographical accounts of six fine scientists are presented, describing how and why they became scientists. These accounts are reproduced with permission from a collection titled *Hundred Reasons to be a Scientist*, brought out by the Abdus Salam International Centre for Theoretical Physics, Trieste, on the occasion of its fortieth anniversary in 2004.

Accounts of the following scientists are presented:

Paul J. Crutzen Atmospheric Science (Netherlands)	(Nobel Prize, Chemistry, 1995)
Vitaly L. Ginzburg Superconductivity (Russia)	(Nobel Prize, Physics, 2003)
Walter Kohn Theoretical Physics (USA)	(Nobel Prize, Chemistry, 1998)
Jean-Marie Lehn Organic Chemistry (France)	(Nobel Prize, Chemistry, 1987)
Abdus Salam Theoretical Physics (UK/Pakistan)	(Nobel Prize, Physics, 1979)
Myriam P. Sarachik Experimental Physics (USA)	(Former President of the American Physical Society)

How I Became a Scientist

Paul J. Crutzen

Max Planck Institute for Chemistry, Mainz,
Germany Scripps Institute of Oceanography, University of
California San Diego, USA

From www.nobelprize.org

I was born in Amsterdam on December 3, 1933, the son of Anna Gurk and Jozef Crutzen. I had one younger sister. My mother's parents moved to the industrial Ruhr region in Germany from East Prussia towards the end of the 19th century. They were of mixed German and Polish origin. In 1929 at the age of 17, my mother moved to Amsterdam to work as a housekeeper. There she met my father. He came from Vaals, a little town in the southeastern corner of the Netherlands, bordering Belgium and Germany and very close to the historical German city of Aachen. He had relatives in the Netherlands, Germany, and Belgium. Thus, from both parents I inherited a cosmopolitan view of the world.

In May 1940 the Netherlands were overrun by the German army. In September of the same year I entered elementary school. My six years of elementary school largely overlapped with the Second World War. Our class had to move several times to different premises in Amsterdam after the German army had confiscated our original school building. The last months of the war, between the Fall of 1944 and Liberation Day on May 5, 1945, were particularly bad. During the cold "hongerwinter" (winter of famine) of 1944-1945 there was a severe lack of food and heating fuels. Also water for drinking, cooking, and washing was available only in limited quantities for a few hours per day, which caused poor hygienic conditions. Many died of hunger and disease, including several of my schoolmates.

In 1946, after successful exam, I entered the "Hogere Burgerschool" (HBS), Higher Citizen School, a five-year long middle school which prepared for University entrance. During those years, chemistry definitely was not one of my favourite subjects. They were mathematics and physics, but I also did well in the three foreign languages: English, French, and German. During my school years I spent considerable time at sports: football, cycling, and my greatest passion, long distance skating on the Dutch canals and lakes. I also played chess. As a child I read especially books about astronomy, geography and discovery expeditions around the world. I was fascinated about the world of the mountains, of which Holland has none, of course. During the war we could not travel anywhere and so I imagined that clouds were snow clad mountains. At home we had a book about Yellowstone National Park in the U.S.A. I read it many times and was fascinated by the pictures. First several years after the war, when I was 17 years old, I saw for the first time real mountains. I met my wife Terttu, a Finnish girl, on a mountain in Switzerland, when I was 21 years old. Yellowstone National Park, together with my family, I visited for the first time in 1975, when I was 42 years old.

Unfortunately, because of a heavy fever, my grades in the final exam of the HBS were not good enough to qualify for a university study stipend. As I did not want to be a further financial burden on my parents for another four years or more (my father, a waiter, was often unemployed; my mother worked in the kitchen of a hospital), I chose to attend the Middelbare Technische School (MTS), Middle Technical School, later called the Higher Technical School (HTS), to train as a civil engineer. Although the MTS took three years, the second year was a practical year during which I earned a modest salary, enough to live on for about two years.

From the summer of 1954 until February 1958, with a 21-month interruption for compulsory military service, I worked at the Bridge Construction Bureau of the City of Amsterdam. After marriage, my wife and I moved to Sweden where I worked in a house construction bureau. However, I was not happy. I longed for an academic career. One day at the beginning of 1959, I saw an advertisement in a Swedish newspaper by the Department of Meteorology of Stockholm University announcing an opening for a computer programmer. Although I had not had the slightest training for such work, I applied for the job and had the great luck to be chosen from among many candidates. On July 1, 1959, we moved to Stockholm and I started with my second profession, that of a computer programmer. The great advantage of being at a university department was that I got the opportunity to follow some of the courses that were offered. By 1963 I could thus fulfil the requirement for the "filosofie kandidat" (corresponding to a Master of Science) degree, combining the subjects mathematics, mathematical statistics, and meteorology.

Around 1965 I was given the task of assisting a scientist from the United States to develop a numerical model of the ozone distribution in the stratosphere, mesosphere, and lower thermosphere. This project got me highly interested in the photochemistry of atmospheric ozone, and I started an intensive study of the scientific literature. This showed me the limited status of scientific knowledge on stratospheric chemistry by the latter half of the 1960's, thus setting the "initial conditions" for my scientific career.

I picked stratospheric ozone, and later atmospheric chemistry and climate studies, as the subject of my research. I was very lucky that there was so much to discover. An important facet of my research was the impact of anthropogenic activities on our atmospheric environment. I discovered that air pollution is not only created by industry and fossil fuel burning, but also by biomass burning in the developing world of the tropics and subtropics. A review of my research is found in my Nobel Lecture[*]. In this choice of research topic I was left totally free. I cannot overstate how I value the generosity and confidence that were conveyed to me by my supervisors Professor Georg Witt, an expert on the aeronomy of the upper atmosphere, and the head of Meteorological Institute, Professor Bert Bolin.

Education, Science and Chance

Vitaly L. Ginzburg

P.N. Lebedev Physical Institute
Moscow, Russia

From www.nobelprize.org

My school years coincided with perhaps the most unfortunate period in the history of Soviet secondary education. Of the old school (gymnasium, etc.) there remained buildings. However, there were several old and supposedly skilled teachers. Chaos reigned over the rest. In 1931 I graduated from a seven-year school, our schooling having been "cut short" by the requirement instituted at that period to learn a proletarian trade in a factory. Finally, a few years later this ill-founded system was changed in favor of one in which schooling went on for 10, and later 11, years.

The absence of the proper "educational" atmosphere, in the family in particular, had the effect that I am under the impression that I gained little from school. Nevertheless, the interest in physics emerged even in those years and a steady one, though I myself do not know why. I was fond of O.D. Khvolson's book "Fizika Nashikh Dnei" (The Physics of our Days), which I read even at school or immediately after graduation, it seems to me. All in all, I never hesitated about going in for physics, but I can recall neither the teacher, nor the textbooks.

On graduation from school, I somehow got fixed up in a job as a laboratory assistant in the Moscow Evening Machine-Building Institute. Initially, I was "in training" in A.A. Bochvar's laboratory of the Institute of Nonferrous Metallurgy and then I found myself in the X-Ray Laboratory. The chiefs were E.F. Bakhmetev and N.K. Kozhina (for some time also Ya.P. Silisskii). The major power was Venya Tsukerman. Leva Al'tshuler was also there. The three of us were on friendly terms and worked together. Of course, I ranked third; the lads were three years older and knew more. Ven'ka called us the 3V's: for Venya, Vitya, and Vladimirovich (that was Al'tshuler's patronimic).

The work in the Laboratory was of benefit to me: it taught me resourcefulness (following Venya's example) and experimental skills. In physics, to say nothing of mathematics, I made no significant progress. The year of 1933 saw the first "free" (i.e. "competitive" rather than by assignment) enrollment to the Moscow State University (MGU), and I decided to enter the Department of Physics. In three months I went formally through the 8th, 9th, and 10th form courses but I am convinced that the lack of a good, regular school had an adverse effect on me. While a schoolboy solves, say, 100 or 1000 problems on trigonometry, logarithms, etc., the number I solved was 10, or 100 times smaller. The same was true of arithmetic. And this told on me forever: I perform calculations badly, slowly, with effort, automatism is lacking. I have always feared and disliked calculations. Of course, behind it is the absence of ability for mathematics (in comparison with the corresponding abilities of the overwhelming majority of fellow-theorists). But this is precisely the reason why the lack of training had so pronounced an effect.

Of course, the lack of regular school adversely affected other aspects, too. At the age of about 30, I read for the first time "Byloe i Dumy" (The Past and Meditation) and many other works of literature (however, I am not sure it is a drawback). Of greater significance is "the Russian language". When I was in my 2nd year at the MGU, all of us took dictation and I made eight mistakes to get "unsatisfactory". Even now I write with mistakes. Making grammatical mistakes is not as significant as the ability for writing, the mastery of style and language. My language is somewhat poor and my phrases are frequently not quite literate. In this connection I recall my conversation with G.S. Gorelik. He had the ability to write well, and to my question "What helps you write so well?" he replied with a question: "How many times a week did you write compositions at school?" I answered, "Something like once a week or once in two weeks, I do not remember". G.S. remarked that he had studied in Switzerland and wrote compositions every day. That is why I still have some gaps in my school knowledge. Disgracefully, I also do not know foreign languages, though, thank God, I have somehow mastered English (but I only can speak, though with mistakes, and make reports, while I am almost unable to write on my own without someone checking it). I am writing all that because I have firmly come to the conclusion that a person needs quite a lot so as to do real work and achieve success and satisfaction. Not knowing languages is, as a matter of fact, a disgrace, to say nothing of the harm to the business. The Europeans do not have such a problem. Any Dutch physicist knows English well and probably also knows German and French: having a facility for languages, one could master a language even without studying it at school—having started from childhood and so forth. But what if a person does not have linguistic abilities?

These are specific abilities indeed. I, for example, am absolutely unable to remember poetry and in general am not able to learn anything by heart (as, for instance, a report). In childhood years at school, I would probably have been able to cope with all that. All my life, I have felt regret that I do not know languages, that I could know more about this and about that. However, when your work is in progress and there are so many interesting things in it, will you learn verbs or the names of constellations? I for one have never been capable of doing that.

All in all, no educational institution would make one into a very good writer, physicist, or mathematician, unless he exhibits the corresponding aptitude. However, first, inclinations alone would not suffice. How many talented people never "realized" their potentialities and what role was played by the shortcomings in education? Second, a good background, training, etc. are supposedly able to make a worthy professional out of a person of average abilities, who would otherwise be a drudge, become a failure, find no satisfaction in work, etc. On the whole, it is all clear. I write wherever I am led by my pen; and this topic has been touched on because I have often pondered over the question as to what losses I have "incurred" due to unfavorable conditions at school. Of course it is impossible to give a clear answer.

On the one hand, as I believe, I was extremely lucky as regards the "realization" of my modest abilities. But, on the other hand, what would have been possible if I had studied in a good ten-year school, to say nothing of "professional" family support (there was none)? Here, I would like to touch upon yet another "favorite" topic which I ponder over quite often. Take for example a sportsman who covered, say, a 100-m distance in 9.9 seconds to become an Olympic champion and a sprinter who did it in 10.2 seconds proved to be the fourth to miss even the bronze medal (the figures are, of course, arbitrary). Here, random circumstances might have played their part: how he had slept, what he had eaten, how he had pushed off the shoe, etc. Fortunately, in science this is not the case: the lot of the fourth is much better, he makes his contribution, writes good papers (with the understanding that the first writes very good ones). But the role of chance and of good luck may still be critical. This is not so for titans like Einstein, for too large is the "safety margin" and the outstripping of others. The talents of Maxwell, Bohr, Planck, Pauli, Fermi, Heisenberg, and Dirac were scarcely dependent as well on the fluctuations of good luck, accidental idea, etc. De Broglie, and even Schrödinger, were, it seems to me, a different matter, to say nothing of numerous Nobel Laureates. M von Laue was a well-qualified physicist, but they state that the idea of X-ray diffraction in crystals was a "beer idea" (Bieridee). Braggs, Roentgen, Zeeman, Stark, Lenard, Josephson, Penzias and Wilson, Hewish and Ryle, Cherenkov, Basov and Prokhorov, as well as 3/4 of the entire list were largely strokes of luck rather than "divine" revelations. I only want to emphasize that chances of success depend both on a lucky strike and a variety of factors, which include health, a timely read article or book, activity, ambition (as a stimulus), and perhaps many other things. An interesting topic.

Finding a Course through Adversity

Walter Kohn

University of California, USA

I was born in 1923 into a middle class Jewish family in Vienna, a few years after the end of World War I, which was disastrous from the Austrian point of view. Both my parents were born in parts of the former Austro-Hungarian Empire, my father in Hodonin, Moravia, my mother in Brody, then in Galicia, Poland, now in the Ukraine. Later they both moved to the capital of Vienna along with their parents. I have no recollection of my father's parents, who died relatively young. My maternal grandparents Rappaport were orthodox Jews who lived a simple life of retirement and, in the case of my grandfather, of prayer and the study of religious texts in a small nearby synagogue, a Schul as it was called. My father carried on a business whose main product was high quality art postcards, mostly based on paintings by contemporary artists commissioned by his firm. The business had flourished in the first two decades of the century but then, in part due to the death of his brother in World War I, to the dismantlement of the Austrian monarchy and to a worldwide economic depression, it gradually fell on hard times in the 1920's and 1930's. My father struggled from crisis to crisis to keep the business going and to support the family. My mother was a highly educated woman with a good knowledge of German, Latin, Polish and French and some acquaintance with Greek, Hebrew and English. I believe that she had completed an academically oriented High School in Galicia. Through her parents we maintained contact with traditional Judaism. At the same time my parents, especially my father, also were a part of the secular artistic and intellectual life of Vienna.

After I had completed a public elementary school, my mother enrolled me in the Akademische Gymnasium, a fine public high school in Vienna's inner city. There, for almost five years, I received an excellent education, strongly oriented toward Latin and Greek. At that time, my favorite subject was Latin, whose architecture and succinctness I loved. By contrast, I had no interest in, nor apparent talent for, mathematics which was routinely taught and gave me the only C in high school. During this time it was my tacit understanding that I would eventually be asked to take over the family business, a prospect which I faced with resignation and without the least enthusiasm.

The Anschluss changed everything: The family business was confiscated but my father was required to continue its management without any compensation; my sister managed to emigrate rather promptly to England; and I was expelled from my school.

In the following fall I was able to enter a Jewish school, the Chajes Gymnasium, where I had two extraordinary teachers: In physics, Dr. Emil Nohel, and in mathematics Dr. Victor Sabbath.

These two inspired teachers conveyed to us their own deep understanding and love of their subjects. Yet again, I take this occasion to record my profound gratitude for their inspiration to which I owe my initial interest in science.

I note with deep gratitude that twice, during the Second World War, after having been separated from my parents who were unable to leave Austria, I was taken into the homes of two wonderful families who had never seen me before: Charles and Eva Hauff in Sussex, England, who also welcomed my older sister, Minna. Charles, like my father, was in art publishing and they had a business relationship. A few years later, Dr. Bruno Mendel and his wife Hertha of Toronto, Canada, took me and my friend Joseph Eisinger into their family. Both of these families strongly encouraged me in my studies, the Hauffs at the East Grinstead County School in Sussex and the Mendels at the University of Toronto. I cannot imagine how I might have become a scientist without their help.

When I arrived in England in August 1939, three weeks before the outbreak of World War II, I had my mind set on becoming a farmer (I had seen too many unemployed intellectuals during the 1930's), and I started out on a training farm in Kent. However, I became seriously ill and physically weak with meningitis, and so in January 1940 my "acting parents", the Hauffs, arranged for me to attend the above-mentioned county school, where—after a period of uncertainty—I concentrated on mathematics, physics and chemistry.

However, in May 1940, shortly after I had turned 17, Churchill ordered most male "enemy aliens" (i.e., holders of enemy passports, like myself) to be interned ("Collar the lot" was his crisp order). I spent about two months in various British camps, including the Isle of Man, where my school sent me the books I needed to study. There I also audited, with little comprehension, some lectures on mathematics and physics, offered by mature interned scientists.

In July 1940, I was shipped on to Quebec City in Canada; and from there, by train, to a camp in Trois Rivieres, which housed both German civilian internees and refugees like myself. Again various internee-taught courses were offered. The one which interested me most was a course on set theory given by the mathematician Dr. Fritz Rothberger and attended by two students. Dr. Rothberger, from Vienna, a most kind and unassuming man, had been an advanced private scholar in Cambridge, England. His love for the intrinsic depth and beauty of mathematics was gradually absorbed by his students.

Later I was moved around among various other camps in Quebec and New Brunswick. Another fellow internee, Dr. A. Heckscher, an art historian, organized a fine camp school for young people like myself, whose education had been interrupted and who prepared to take official Canadian High School exams. In this way I passed the McGill University junior Matriculation exam and exams in mathematics, physics and chemistry on the senior matriculation level. At this point, at age 18, I was pretty firmly looking forward to a career in physics, with a strong secondary interest in mathematics.

I mention with gratitude that camp educational programs received support from the Canadian Red Cross and Jewish Canadian philanthropic sources. I also mention that in most camps we had the opportunity to work as lumberjacks and earn 20 cents per day. With this princely sum, carefully saved up, I was able to buy Hardy's Pure Mathematics and Slater's Chemical Physics, books which are still on my shelves. In January 1942, I was released from internment and welcomed by the family of Professor Bruno Mendel in Toronto. At this point I planned to take up engineering rather than physics, in order to be able to support my parents after the war. The Mendels introduced me to Professor Leopold Infeld who had come to Toronto after several years with Einstein. Infeld, after talking with me (in a kind of drawing room oral exam), concluded that my real love was physics and advised me to major in an excellent, very stiff program, then called mathematics and physics, at the University of Toronto. He argued that this program would enable me to earn a decent living at least as well as an engineering program.

I was fortunate to find an extraordinary mathematics and applied mathematics program in Toronto. Luminous members whom I recall with special vividness were the algebraist Richard Brauer, the non-Euclidean geometer, H.S.M. Coxeter, the aforementioned Leopold Infeld, and the classical applied mathematicians John Lighton Synge and Alexander Weinstein. This group had been largely assembled by Dean Beatty. In those years the University of Toronto team of mathematics students, competing with teams from the leading North-American Institutions, consistently won the annual Putman competition. (For the record I remark that I never participated). Physics too had many distinguished faculty members, largely recruited by John C. McLennan, one of the earliest low temperature physicists, who had died before I arrived. They included the Raman specialist H.L. Welsh, M.F. Crawford in optics and the low-temperature physicists H.G. Smith and A.D. Misener. Among my fellow students was Arthur Schawlow, who later was to share the Nobel Prize for the development of the laser.

During one or two summers, as well as part-time during the school year, I worked for a small Canadian company which developed electrical instruments for military planes. A little later I spent two summers, working for a geophysicist, looking for (and finding!) gold deposits in northern Ontario and Quebec.

After my junior year I joined the Canadian Army. An excellent upper division course in mechanics by A. Weinstein had introduced me to the dynamics of tops and gyroscopes. While in the army I used my spare time to develop new strict bounds on the precession of heavy, symmetrical tops. This paper, "Contour Integration in the Theory of the Spherical Pendulum and the Heavy Symmetrical Top" was published in the Transactions of American Mathematical Society. At the end of one year's army service, having completed only 2 1/2 out of the 4-year undergraduate program, I received a war-time bachelor's degree "on active service" in applied mathematics.

In the year 1945-6, after my discharge from the army, I took an excellent crash master's program, including some of the senior courses which I had missed, graduate courses, a master's thesis consisting of my paper on tops and a paper on scaling of atomic wave- functions.

My teachers wisely insisted that I do not stay on in Toronto for a Ph.D, but financial support for further study was very hard to come by. Eventually I was thrilled to receive a fine Lehman fellowship at Harvard. Leopold Infeld recommended that I should try to be accepted by Julian Schwinger, whom he knew and who, still in his 20's, was already one of the most exciting theoretical physicists in the world.

Arriving from the relatively isolated University of Toronto and finding myself at the illustrious Harvard, where many faculty and graduate students had just come back from doing brilliant war-related work at Los Alamos, the MIT Radiation Laboratory, etc., I felt very insecure and set as my goal survival for at least one year. The Department Chair, J.H. Van Vleck, was very kind and referred to me as the Toronto-Kohn to distinguish me from another person who, I gathered, had caused some trouble. Once Van Vleck told me of an idea in the band-theory of solids, later known as the quantum defect method, and asked me if I would like to work on it. I asked for time to consider. When I returned a few days later, without in the least grasping his idea, I thanked him for the opportunity but explained that, while I did not yet know in what subfield of physics I wanted to do my thesis, I was sure it would not be in solid state physics. This problem then became the thesis of Thomas Kuhn (later a renowned philosopher of science), and was further developed by myself and others. In spite of my original disconnect with Van Vleck, solid state physics soon became the center of my professional life and Van Vleck and I became lifelong friends.

After my encounter with Van Vleck I presented myself to Julian Schwinger requesting to be accepted as one of his thesis students. His evident brilliance as a researcher and as a lecturer in advanced graduate courses (such as waveguides and nuclear physics) attracted large numbers of students, including many who had returned to their studies after spending "time out" on various war-related projects.

I told Schwinger briefly of my very modest efforts using variational principles. He himself had developed brilliant new Green's function variational principles during the war for waveguides, optics and nuclear physics (soon afterwards Green's functions played an important role in his Nobel-Prize-winning work on quantum electrodynamics). He accepted me within minutes as one of his approximately 10 thesis students. He suggested that I should try to develop a Green's function variational method for *three*-body scattering problems, like low- energy neutron-deuteron scattering, while warning me ominously, that he himself had tried and failed. Some six months later, when I had obtained some partial, very unsatisfactory results, I looked for alternative approaches and soon found a rather elementary formulation, later known as Kohn's variational principle for scattering, and useful for nuclear, atomic and molecular problems. Since I had circumvented Schwinger's beloved Green's functions, I felt that he was very disappointed. Nevertheless he accepted this work as my thesis in 1948.

Looking back I feel very fortunate to have had a small part in the great drama of scientific progress, and most thankful to all those, including family, kindly "acting parents", teachers, colleagues, students, and collaborators of all ages, who made it all possible. It has been an interesting journey.

Walk with Responsibility

Jean-Marie P. Lehn

Laboratoire de chimie supramoléculaire Strasbourg, France

(From www.nobelprize.org)

Science offers most exciting perspectives for the future generations. It promises a much more complete understanding of the universe, an always greater creative power of chemical sciences over the structure and transformations of the inanimate as well as of the living world, an increasing ability to take control over disease, aging and even over the evolution of the human species, a deeper penetration into the working of the brain, the nature of consciousness and the origin of thought.

First, I would like to say a few words about my own field of activity, chemistry, about what is making it so attractive to me.

Indeed, chemistry plays a central role by its place in the natural sciences and in knowledge, as well as by its economic importance and omnipresence in our everyday lives. Being present everywhere, it tends to be forgotten and to go unnoticed. It does not advertise itself but, without it, those achievements we consider spectacular would not see the light of day: therapeutic exploits, feats in space, marvels of technology, and so forth. It contributes to meeting humanity's needs in food and medication, in clothing and shelter, in energy and raw materials, in transport and communications. It supplies materials for physics and industry, models and substrates for biology and medicine, properties and processes for science and technology.

In addition to the exploration of the molecules of life, chemistry seeks non-natural species, possessing a desired chemical or physical property. It opens wide the door to the creative imagination of the chemist at the meeting point of chemistry with biology and physics.

Like the artist, the chemist engraves into matter the products of creative imagination. The stone, the sounds and the words do not contain the works that the sculptor, the composer, the writer express from them. Similarly, the chemist creates original molecules, new materials and novel properties from the elements provided by nature.

The essence of chemistry is not only to discover but to invent and above all to create. The book of chemistry is not only to be read but to be written! The score of chemistry is not only to be played but to be composed!

Beyond the general progress of knowledge and the technological development, the most important impact science can and must have on society is the *spirit* that it implies, the scientific, rational approach towards the world, life and society.

Science education in our schools, colleges and universities as well as for the general public must be a major priority, so as

- to train the researchers, discoverers and creators of tomorrow;
- to lift irrational fears and rejections;

- to develop the scientific spirit, the scientific attitude, in order to fight the obscure, the deceitful, the irrational.

A continuing and aggravating problem is the unacceptable *North/South imbalance*. It is the responsibility of the developed countries to offer solutions. Again scientific progress is crucial. One may hope that progress in medicine and advanced technologies resulting from research in the developed countries, provide means for fighting disease and allowing sustainable development in the less advanced countries.

A very actual issue concerns the attitude of the scientist with respect to *ethics and society*. It is my strong opinion that the scientist is first of all responsible towards the truth. Ethics is a relative notion and ethical evaluations change with time, location and progress of knowledge. Pursuit of knowledge and truth must supersede present considerations on nature, life or the world, for our vision of today can only be a narrow one and we have no right to switch out the lights of the future.

These perspectives for the future of science, for *our* future, have already been expressed in most fitting terms by this quintessence of the artist-scientist, *Leonardo da Vinci* when he wrote:

"Where nature finishes to produce its own species, man begins, using natural things, in harmony with this very nature, to create an infinity of species".

The future of science and of mankind lies in the hands of the coming generations. May they take up the challenge. Prometheus has conquered the fire and we cannot give it back. We have to walk, with enthusiasm, determination and a deep sense of responsibility, the way from the tree of knowledge to the control of destiny.

Science and Scientists in Developing Countries

Abdus Salam

International Centre for Theoretical Physics,
Trieste, Italy

(From www.nobelprize.org)

I was born in the country town of Jhang, then part of British India, now Pakistan, in 1926. My father was a teacher and educational official in the Department of Education and my mother was a housewife. I had six brothers and one sister. My family was by no means rich. My father took a vast amount of interest in my school work. He had great ambitions for me. I was destined for the Indian Civil Service, entry to which was by competitive examination. However, this was not to be—as events in my life took a different turning.

When I was at school in about 1936 I remember the teacher giving us a lecture on the basic forces in Nature. He began with gravity. Of course we had all heard of gravity. Then he went on to say "Electricity. Now there is a force called electricity, but it doesn't live in our town Jhang, it lives in the capital town of Lahore, 100 miles to the east". He had just heard of the nuclear force and he said "that only exists in Europe". This is to demonstrate what it was like to be taught in a developing country.

When I was 14, I won a scholarship to Government College, Lahore, with the highest marks ever recorded. I recall that when I cycled home from Lahore, the whole town turned out to welcome me. I wrote my first research paper when I was about sixteen years of age which was published in a mathematics journal but I wasn't actually hooked on research till I went to Cambridge University.

I was very fortunate to get a scholarship to go to Cambridge. The famous Indian Civil Service examinations had been suspended because of the war and there was a fund of money that had been collected by the Prime Minister of Punjab. This money had been intended for use during the war, but there was some of it left unused and five scholarships were created for study abroad. It was 1946 and I managed to get a place in one of the boats that were full with British families who were leaving before Indian Independence. If I had not gone that year, I wouldn't have been able to go to Cambridge; in the following year there was the partition between India and Pakistan and the scholarships simply disappeared.

At Cambridge, I achieved a First in the Mathematics Tripos in two years. I still had a third year free in the sense that I had the scholarship and the choice of whether to go on with higher mathematics—that's part III of the mathematics tripos—or to do the physics tripos. On the advice of my tutor, Fred Hoyle, who said "If you want to become a physicist, even a theoretical physicist, you must do the experimental course at the Cavendish. Otherwise, you will never be able to look an experimental physicist

in the eye", I joined the Cavendish Laboratory where Rutherford
had carried out his experiments on the structure of the atom.
The Cavendish was an outstanding laboratory for experimental
work and a focus for physicists around the world. However, I had
very little patience with experimental equipment. To be a good
experimenter you must have patience towards things which are
not always in your control. I think a theoretician has got to be
patient too, but that is with something of his own creation, his
own constructs, his own stupidities.

The very first experiment I was asked to do was to measure
the difference in wave length of the two sodium D lines, the
most prominent lines in the sodium spectrum. I reckoned that
if I drew a straight line on the graph paper then its intercept
would give me the required quantity I wanted to measure.
Mathematically, a straight line is defined by two points and
if you take one other reading then mathematically that should
be enough since you then have three points on that line, two
to define the straight line and the third one to confirm. I spent
three days in setting up that equipment. After that I took
three readings, and took them to be marked. In those days the
marking of experimental work in the class counted towards
your final examination. Sir Denys Wilkinson was one of the
men who supervised our experimental work, and I took it to
him. He looked at my straight line, and asked "What's your
background?" I said "Mathematics". He said "Ah, I thought so.
You realise that instead of three readings you should have taken
one thousand readings and drawn a straight line through them".
I had by that time dismantled my stuff and didn't want to go
back. So I tried very hard to avoid Denys Wilkinson during
the rest of the year. I still remember the results came out in
1949. I was looking at the results sheets hung in the Cavendish
and Wilkinson came up behind me. He looked at me and said
"What sort of class have you got?" and I very modestly said
"Well, I've got a first class". He turned full circle on his heel,

three hundred and sixty degrees, turned completely round, and said "Shows you how wrong you can be about people".

I went back to Lahore in 1951 and taught there at the University. But as a physicist, I was completely isolated. It was very difficult to get the journals and keep in touch with my subject. I had to leave my country to remain a physicist. Now, it is the lack of this contact with others that is the biggest curse of being a scientist in a developing country. You simply do not have the funds, the opportunities, which those from richer countries enjoy as a matter of course. There are not the communities of people thinking and working in the same fields. This is what we have tried to cure by bringing people together at the International Centre for Theoretical Physics which I founded in Trieste in 1964. The Centre provides the possibility for scientists to remain in their own country for the bulk of the time, but come to the Centre to carry out research for three months or so. They meet people working in the same subject, learn new ideas and can return to their own country charged with a mission to change the image of science and technology in their own country.

I returned to Cambridge in 1954 as a lecturer and Fellow of St. John's College. Three years later, I accepted a professorship at Imperial College, London, where I succeeded in establishing one of the best theoretical physics groups in the world.

The pinnacle of my physics career came in 1979 when I shared the Nobel Physics Prize with Sheldon Glashow and Steven Weinberg for our unification of electromagnetism and the weak nuclear force in the 'electroweak' (a word which I invented in 1978) theory, one of the major achievements of twentieth century physics. This theory had made predictions that could be verified by experiment. The most revealing of these was that a new particle exists at extreme energies. To test this theory we had to convince the experimental physicists working on the great particle accelerators to build new equipment: To create, in principle, conditions that would be similar to those first few moments in the birth of the universe. In 1983 the final confirmation was obtained with the discovery that the predicted particles—the intermediate vector bosons—did exist. Called W^+, W^- and Z^0, these hypothetical particles were seen for a few fleeting moments under the cosmic conditions of the CERN accelerator. This temporary existence was enough to demonstrate that the unification theory was an accurate description of the fundamental nature of matter. This experimental verification led to the award of the Nobel Prize to Carlo Rubbia and Simon van der Meer in 1984.

I spoke earlier of the difficulties of doing science in developing countries. I would like to conclude with an appeal. Funds allotted for science in developing countries are small, and the scientific communities sub-critical. Developing countries must realize that the scientific men and women are a precious asset. They must be given opportunities, responsibilities for the scientific and technological developments in their countries. Quite often, the small numbers that exist are underutilized. The goal must be to increase their numbers because a world divided between the haves and have-nots of science and technology cannot endure in equilibrium. It is our duty to redress this inequity.

Why Physics?

Myriam P. Sarachik

City College of New York, USA

(From www.sci.ccny.cuny.edu)

Owing to historical circumstances, my early years were eventful and quite unusual. I was born in 1933, the year Hitler came to power, and my life's trajectory was irrevocably altered by subsequent events which caused my family to flee Belgium. We found refuge in late 1941 in Cuba, where I grew up from age 8 to 13 and a half, and then immigrated to the United States, the land of promise, freedom and limitless possibilities. This is where I attended high school, then College and graduate school, where I obtained my advanced degrees.

My family and my community were quite traditional. Women were not expected to work outside the home. Women were responsible for raising and caring for children, which is certainly very hard work. But they did not embark on their own career. If a woman worked outside the home, it was a sign that her husband was an inadequate provider.

I remember my very early years as moderately boring and depressing. My life changed on the day I entered school. It opened an entirely new and exciting world for me. Reading was a pleasure. Numbers were a pleasure. I truly loved it! Before I started school, I was considered sweet, a good child, but a bit of a sad sack. Now I was considered worthy and special. After first grade, my education was interrupted for more than a year as we fled from country to country. Although my father disapproved spending money on books (we were in terrible danger and there was no time or money for such things), my mother bought books for me to read. And I read each book over and over and over again until I got another book to read. I was eight years old when my family arrived in Havana, and I could then go to school again. I was interested in many things. I liked grammar, languages, anatomy, geography, almost everything. I read voraciously, I played the piano and loved music passionately, and I took special pleasure in arithmetic, algebra, patterns and quantitative things generally. What made me finally settle on physics when I had to make the choice? I'm not sure I know the answer. I found physics the hardest subject I had ever encountered, and I did quite badly at it in the beginning. It was a real challenge. Other things had been relatively easy for me. Physics was TOUGH and physics was held in very high regard. It was considered the epitome of intellectual achievement by my father, whom I loved dearly, and respected and admired very highly. My father was curiously conflicted about my scholastic pursuits. On the one hand, he placed great value on intellectual excellence and encouraged me to excel; he surely would have chosen to be a physicist had life offered him that chance. On the other hand, I was a girl, and girls should marry and have children.

I earned my Bachelor's Degree in 1954 with a major in physics (at Barnard College, although all the courses were given across the street at Columbia because there weren't enough of us girls interested in the subject). Philip Sarachik and I were married that summer, and it was time to move on to the life I was expected to live. But there was no harm in waiting a bit, and I took a job at IBM Watson Laboratories very near Columbia University. By this time, I had been quite captured and captivated by physics. It was still tough but I could do it, and it was fascinating. I very much wanted to go on for a Ph.D., but I felt I must not. But, there was no harm in taking one or two graduate courses down the street at Columbia, was there? My husband then decided to work toward a Ph.D. in Electrical Engineering. I then allowed myself to do the same.

There were many challenges waiting for me along my chosen path. There were very few women candidates for Ph.D.'s in physics in those days. I received no encouragement from the faculty; on the other hand, except for many faculty members' negative perceptions of women's ability and role, there were no overt blocks. We all had to pass the same exams, we all had to do a thesis, and we all had to defend that thesis.

The next step on the road was quite different: I had an incredibly tough time landing my first job. I believe that part, perhaps a large part, of the problem was that I had become a mother, having given birth to my first daughter Karen. But I persisted mightily, and I managed to meet that challenge too.

I will not dwell on my middle years. In brief, after doing a postdoc at IBM Watson Laboratories, and another at Bell Telephone Laboratories, I joined the faculty at City College of New York, where I have spent essentially my entire professional life. I have enjoyed every aspect of my work at CCNY. I began as a middling teacher and grew to love teaching, learning to do it better and better with time. I've taken on the usual responsibilities in my department and university over the years.

But I've derived the greatest joy and satisfaction from my research, where I believe I've made a few significant contributions. While a postdoc at Bell Labs, I did experimental measurements that established a one-to-one correspondence between the presence of a localized moment and the occurrence of a minimum in the resistance versus temperature in certain alloys; contemporaneously, Jun Kondo performed a now-famous calculation that demonstrated that the minimum is indeed due to a local moment. This solved a long-standing puzzle that had existed since the 1930's. More recently, I have done work (with Sergey Kravchenko) that raises the possibility that an unexpected metallic state can exist in two dimensions. And in 1996, a Ph.D. student in my group, Jonathan Friedman, and I discovered tunneling of a large magnetic moment through the observation of steps in the magnetization curve of a molecular nanomagnet, Mn12-acetate, a discovery that has stimulated an enormous amount of activity in what's now referred to as "single molecule magnets".

My interests have taken me to issues and places outside City College. I have participated in many efforts to defend the human rights of scientists. I have served the physics community in a number of ways: through advisory assignments, service to the American Physical Society, organizing conferences, and so on. These activities culminated in my election to the position of President of the American Physical Society in 2003, a year that was unique for its intensity, involvement, and rewards.

My life as a physicist has been enormously satisfying and great fun. That doesn't mean that every moment has been fun. There have been problems and challenges along the way, and there have been setbacks, small and large. One of the most exhilarating aspects of being a physicist (or a mathematician, chemist, biologist ...) is that one continues to learn, stretch and expand. I'm referring not only to finding truly new facts and phenomena through doing research. I'm referring also to the joy of learning things that are known and understood by others, but that you now understand for the first time. It's a wonderful challenge.

2. WHAT IS SCIENCE ALL ABOUT?

Is science something special? What is unique about science compared to other intellectual endeavours? What are the special features of science, if any? These are some of the questions that young people and those who do not have a background in science may ask. First of all, let me state that science, like many other areas, has been pursued for long because of the creative urge of human beings. Let us not forget that centuries ago in India and Europe, many important discoveries were made. Science will continue to survive and prosper as long as there are human beings. An aspect that distinguishes science from other scholarly pursuits is that much of science is based on experimentation.

I like what G.N. Lewis wrote about science: 'The strength of science lies in its naïveté. Science is like life itself. If we could foresee all the obstacles that lie in our path we would not attack even the first, but would settle down to self-centred contemplation. The average scientist, unequipped with the powerful lenses of philosophy, is a near-sighted creature, and cheerfully attacks each difficulty in the hope that it may prove to be the last . . . I take it that the scientific method of which so much has been heard, is hardly more than the native method of solving problems, a little clarified from prejudice and a little cultivated by training . . . I have no patience with attempts to identify science with measurement, which is but one of its tools, or with any definition of a scientist which would exclude a Darwin, a Pasteur, or a Kekulé . . . The scientist is a practical man and his are practical aims. He does not seek the ultimate, but the proximate. The theory that there is an ultimate truth, although very generally held by mankind, does not seem useful to science except in the sense of a horizon toward which we may proceed.'

Some features of science

Imagination plays a crucial part in science. However, as Feynman stated, 'Whatever we are allowed to imagine in science has to be consistent with everything else that we know. We cannot allow ourselves to imagine things which are obviously in contradiction to the known laws of nature. One has to have the imagination to think of something that has never been seen or

Richard P. Feynman

(From www.nobelprize.org)

heard before. At the same time, the thoughts are restricted by the conditions that come from our knowledge of the way nature really is. The problem of creating something which is new, but which is consistent with everything which has been seen before, is one of extreme difficulty.' The difficulty with science is often not with new ideas, but in escaping the old ones. A certain amount of irreverence seems to be essential for the creative pursuit of science.

Science does not allow exceptions. Without this feature, there would be no determinism in science or there would be no science at all. Another interesting feature is that 'full explanations are often seized in their essence by the scientist long in advance of any possible proof'.

Unanimity of opinion is not necessary in science. While unanimity of opinion may fit a religious organization, a variety of opinions is essential in science. Science is the only area which has not produced sects (although there are disciplines and schools in science). It is founded on analysis and synthesis. Importantly, the same method of science is followed around the world. In questions of science, as Galileo said long ago, 'the authority of a thousand is not worth the humble reasoning of a single individual'. It is useful to remember the motto of the Royal Society, *'Nullius in Verba'* (we do not take anybody's word for it). Jacob Bronowski has written, 'By the worldly standards of public life, all scholars in their work are, of course, oddly virtuous. They do not make wild claims, they do not cheat, they do not try to persuade at any cost, they appeal neither to prejudice nor to authority, they are often frank about their ignorance, their disputes are fairly decorous, they do not confuse what is being argued with race, politics, sex or age, they listen patiently to the young and to the old, who both know everything. These are the great virtues of scholarship and they are particularly the virtues of science'.

'Science has nothing to be ashamed of, not even in the ruins of Nagasaki.'

Experimental science

Pyotr Kapitza stated, 'theory is a good thing, but a good experiment lasts forever'. Leonardo da Vinci said that 'there is no higher or lower knowledge but only one flowing out of experimentation.' According to Eigen, 'A theory has the only alternative of being right or wrong, a model has a third possibility. It may be right, but irrelevant.' Theoretical research without relation to experimentation tends to become sterile. I like what Michael Faraday said about experiments. 'Nothing is too wonderful to be true if it be consistent with the laws of nature and in such things as these, experiment is the best test of such consistency.' Faraday is probably the greatest experimentalist of all time.

While it is pointless to discuss the relative importance of experimental and theoretical research, there seems little doubt that when experimental research dies, science also dies in that country. This should be a matter of concern to us, since there are very few people doing good experimental work in India, at least in physical sciences. What often becomes essential is to find theoretical explanations for experimental observations. Experiment and theory are both necessary to understand a phenomenon or a property.

Science and beauty

Aesthetic criteria have often been invoked in science and it has been said that the subject has something intrinsically poetic. Science has been considered a meeting place of two kinds of poetry—the poetry of action and the poetry of thought. According to Niels Bohr 'when it comes to atoms, language can be used only as in poetry. The poet, too, is not merely so concerned with describing facts as with creating images.'

According to G.H. Hardy, 'beauty is the first test; there is no permanent place in the world for ugly mathematics.' The place of beauty in equations was stressed by Paul Dirac. According to him, 'It is more important to have beauty in one's equations than to have them fit experiment. If one is working from the point of view of getting beauty in one's equations, and if one has really a sound insight, one is on a sure line of progress. If there is not complete agreement between the results of one's work and experiment, one should not allow oneself to be too discouraged, because the discrepancy may well be due to minor features that are not properly taken into account and that will get cleared up with further developments of theory.' When Einstein was reproached that his formula

G.H. Hardy

of gravitation was longer and more cumbersome than that of Newton, he apparently said that 'if one were to describe the truth, leave elegance to the tailor'. Yet, Einstein's equation has been considered by many to be complete and beautiful.

Profs. C.N.R. Rao, M.G.K. Menon and S. Ramaseshan with Prof. S. Chandrasekhar at the golden jubilee meeting of the Indian Academy of Sciences (1984)

Beauty in science has found excellent expression in the writings of the astrophysicist, S. Chandrasekhar, who quotes Henri Poincaré. 'The scientist does not study Nature because it is useful to do so. He studies it because he takes pleasure in it; and he takes pleasure in it because it is beautiful. If nature were not beautiful, it would not be worth knowing and life would not be worth living . . . I mean the intimate beauty which comes from the harmonious order of its parts and which a pure intelligence can grasp. . . . It is because simplicity and vastness are both beautiful that we seek by preference, simple facts and vast facts; that we take delight, in following the giant courses of the stars, in scrutinizing with a microscope that prodigious smallness which is also a vastness, and, in seeking in geological ages the traces of the past that attracts us because of its remoteness.' Commenting on Poincaré's perception,

J.W.N. Sullivan states that 'it is in the aesthetic value that the justification of the scientific theory is to be found and with it the justification of the scientific method . . . The measure in which science falls short of art is the measure in which it is incomplete as science.'

I have some misgivings about taking arguments based on aesthetic factors too literally since they may not be conducive to the vigorous research demanded today. After all, we are no longer in the golden age when one was mainly concerned about the universal physical laws for all time. Furthermore, to limit beauty to mathematical equations alone would be unfair to science. There is beauty in the architecture of molecules and materials, as well as in the pathways of transformations. There is also beauty in properly designed and well-executed experiments. There is beauty even in the way science works and develops. As Faraday stated, 'It is the great beauty of our science that advancement in it, whether in a degree great or small, instead of exhausting the subject of research, opens the doors to further and more abundant knowledge, overflowing with beauty and utility.'

Science and utility

The basic urge to do science is not for utilitarian purposes, although science has its uses. Hermann von Helmholtz stated that 'whosoever in the pursuit of science seeks after immediate practical utility, be rest assured that he seeks in vain'. There is some truth in it because while pursuing science it would be difficult to fully imagine or to expect the full implications or usefulness of the results, although significant spin-offs may indeed result later. Hardy, talking about his work, stated, 'I have never done anything useful, no discovery of mine has made, or is likely to make directly or indirectly, for good or ill, the least difference to the amenity of the world. Judged by all practical standards, the value of my mathematical life is nil; and outside mathematics, it is trivial anyhow. I have just one chance of escaping a verdict of complete triviality, that I may be judged to have created something is undeniable; the question is about its value'.

While science is pursued by good scientists without any expectation, one cannot forget the immense benefits obtained through science. It is said that when William Ewart Gladstone, the Chancellor of the Exchequer, asked Faraday about the practical worth of electricity, Faraday is said to have replied, 'One day, sir, you may tax it.' As scientists, we should not undermine the so-called applied science. Applied science can be as challenging as pure science. Scientists indulging in fundamental research tend to have a poor opinion of those who do applied science and this attitude is not helpful. George Porter stated, 'There is science that has been applied and science that is yet to be applied.'

Science requires many types of workers. To quote Peter Medawar, 'Among scientists are collectors, classifiers and compulsive tidiers-up; many are detectives by temperament and many are explorers; some are artists and others are artisans. There are poet–scientists and philosopher–scientists and even a few mystics.' But one of the features that characterizes modern scientific research is that 'it is the art of the solvable', unlike politics which is the 'art of the possible'. 'Both are immensely practical-minded. Good scientists study the important problems they think they can solve.' Pandit Nehru had the firm belief that 'it is science alone that can solve the problems of hunger and poverty, insanitation and illiteracy, of superstition and deadening custom and tradition, of vast resources running to waste, of a rich country inhabited by starving people . . . The future belongs to science and to those who make friends with science.'

Science has been a great instrument of social change, all the greater because its object is not change, but knowledge. Its silent appropriation of this important function is the most vital of all the revolutions associated with the development of modern civilization. Let us not forget that the discovery of the transistor created a greater social revolution than any other idea that one has propounded.

Nature's revenge

Science is changing all the time, and the world is also changing in many ways. Many of the problems and challenges that the world faces today are indirectly or directly related to science. Let me give a typical example. Chemists discovered polymers and the discovery has been a great boon to the world. The plastics which have helped mankind in many ways have become an environmental risk today. The piles of plastic bags along the roads of India have had devastating effects on nature. When laptops and email came into vogue in a big way, people promised President Clinton that there would be little use of paper in the future. Everything would be done electronically through computers, and even newspapers would be electronic. What has actually happened, however, is that we use much more paper today than ever before. We can see nature's revenge in many ways. Scientists discovered antibiotics and many other drugs, which have saved human kind, but today many people exhibit drug resistance. The cars we adored when we were young and the aeroplanes that we have enjoyed all these years have become the source of pollution and we have to find alternative fuels to save the atmosphere. We can, therefore, never claim that we have done all the science (research) that is required and found solutions to problems of mankind. There will be newer problems faced by mankind in the future, many of which may be the result of applications of science and technology. One thing is for certain, science will be required to solve these problems as well.

Support for science

In advanced countries, science is supported partly through sources other than those of the government. Even in these countries, state intervention has been essential to ensure continued, balanced support to accrue the benefits of scientific work for national development. In countries like India, there is no source other than the government for the support of science. Unfortunately, there is little appreciation in many quarters regarding the responsibility of the government in ensuring that science is well supported and that the health of the subject is satisfactory. This is because many of the politicians, planners and administrators are ignorant of science and this defect runs through the civil service. It is said to be nearly universal in the House of Commons in Britain, and is shared by the general public, including a large proportion of those engaged in industrial and commercial enterprise. This ignorance is affecting science badly, especially in countries such as India. I am reminded of the story narrated by Al-Biruni. Once a sage was asked why scholars and scientists always flocked to the doors of the rich, while the rich are not inclined to call at the doors of the scholars. The sage answered, 'The scholars are well aware of the use of money, but the rich are ignorant of the nobility of science.'

Frederick Dainton stated some time ago, 'Accountants and second-rate business school jargon are in the ascendant. Costs, which rise rapidly, and are easily ascertained and comprehensible, now weigh more heavily in the scales than the unquantifiable and unpredictable values, and future material progress. Perhaps, science will regain its lost primacy as people and the government begin to recognize that sound scientific work is the only secure basis for the construction of policies and to ensure the survival of mankind without irreversible damage to planet Earth.'

Scientific temper

It is not an exaggeration if one were to state that science has become an essential component of everything human beings do and aspire for. Science has entered our lives, and scientific terms (e.g., computer, antibiotic, nano, IT) have entered common vocabulary. Knowledge of science at an appropriate level is required to converse intelligently and purposefully. The status of a country is indirectly determined by the status that science occupies there. Science is required in every country, however small or poor, and by every human being, independent of their geographical location or economic status. Furthermore, decision making in society and governments can best be done if there is a scientific approach. Scientific temper is essential to tackle problems in society and in life itself. It is of utmost importance that the inculcation of scientific temper among citizens gets the necessary importance in the plans of governments and other organizations. This is especially important in developing countries, in order to eliminate disadvantages arising from superstitions and obscurantism. One of my friends recently told me, 'You know, many scientists do not have a scientific attitude.' This summarizes the problem nicely.

Science as a way of life

Science has been an essential part of my life. When I started working in science, I did not know that it would play such an intrinsic role in my life for six decades. I took up science as a profession and as I gradually went on working, I found that science was not only running my life but was dictating its terms. There was no way I could plan my life without science. If my work in the laboratory gets affected by what I do in other spheres of life, I stop indulging in those activities. I have found that doing science alone is not as interesting as doing science with young people. Being a teacher has, therefore, been extremely important to me. Being a teacher alone without doing research has no meaning either. The great Sanskrit poet Kalidasa wrote many centuries ago, 'If a professor thinks what matters most is to have gained an academic post, where he can earn a livelihood, and then neglect research, let controversy rest, he is but a petty tradesman at best, selling retail the work of other men.'

Those of us who pursue science by working with young people cannot forget our primary responsibility in bringing up new generations of scientists. We have to give everything we have to the young and guide them in a manner that they are able to blossom as scientists, fully exploiting their native capabilities. It is difficult to convey the excitement in science unless one has experienced 'the dramatic feeling of sudden enlightenment that floods the mind when the right idea finally clinches into space'. By the way one lives and by the way one works, it is possible that one may be able to communicate the right message to the young.

A great diversion to many a good scientist has been their getting involved in administration. Unfortunately, it is a necessary and an unavoidable evil. I realize that one cannot possibly be like Michael Faraday. Faraday, when approached to take up the Presidentship of the Royal Society, apparently wrote, 'I must remain plain Michael Faraday to the last; and let me now tell you that if I accepted the honour which the Royal Society desires to confer upon me, I would not answer for the integrity of my intellect for a single year.' It is too much to ask of ordinary mortals like myself, yet there is no doubt that those of us who have done even a little bit of administration have lost valuable time for research. Those of us who do research beyond all other commitments, often find that we occasionally miss something crucial and important because of our preoccupation with administration. Administrative chores and worries come in the way of the crucial minutes for unperturbed meditation. I sincerely advise all those aspiring young scientists who really want to make a mark in science, not to touch any form of administration unless it is managing a small research group.

Salary and status are irrelevant parameters for a scientist. I have felt that I have been an overpaid research scholar doing what I like most. It so happens that scientists themselves treat administrators with undue respect and prefer politicians to inaugurate symposia. Some of us suffer from a persecution complex, possibly with justification. A man does not attain the status of Galileo because he is persecuted. He must also be a good scientist. It is necessary that we scientists have a good sense of humour.

Some scientists are conservative about publishing, while others are prolific. While there is no simple rule about how much to publish or not to publish, there is no denying the fact that, as Benjamin Franklin as well as Faraday put it, the main activity of scientists is to 'work, finish and publish'. I am not ashamed of publishing. I have known great scientists who have published several hundred papers. Both Faraday and C.V. Raman published over 400 papers. I have also known extraordinary men who have published very few papers. I have known theoreticians, like Nevill Mott, who have published consistently and constantly for several decades. The real problem in India, however, is that many of us publish a large number of papers which are not significant. A small number of good papers coming from a large country like India has little visibility.

Patrick Blackett was of the view that 'A first rate laboratory is one where even mediocre scientists produce outstanding work.' This is probably what happens when some of us go to more advanced countries. We must have at least some good laboratories in India where we can perform and where most of us can do our best to produce results that are worthwhile. While facilities and support may be declining, science itself is galloping. The pace at which science is moving is so great indeed that the frontiers of today often become commonplace science tomorrow. The current rate of progress in science allows no respite and competition is intense in most areas. Safe problems are easier to work on, but get little notice. To be counted, one has to fearlessly take up difficult, contemporary problems.

Freedom may be the first born daughter of science. Unfortunately, science seems to have lost her freedom in recent years. 'It has become a productive force' as Kapitza said. 'It has even become rich, but has become enslaved.' I am not sure whether Faraday, Rutherford and Raman would be happy to do science in the present circumstances. Public opinion is not discriminating and is likely to interpret demands of scientists as meaning that science can be had for money. But science is not for sale.

Samuel Coleridge wrote long ago, 'the first man of science was he who looked into the thing, not to learn whether it furnished him with food, or shelter, or weapons, or tools, armaments or playwiths, but who sought to know it for the gratification of knowing.' I believe that this spirit is still prevalent among some of us. Raman said that 'the pursuit of science derives its motive power from what is essentially a creative urge. The painter, the sculptor, the architect and the poet, each in his own way, derives his inspiration from nature and seeks to represent her through his chosen medium. . . . The man of science, like the exponents of art, subjects himself to a rigorous discipline . . . Science is a fusion of man's aesthetic and intellectual functions devoted to the representation of nature.'

I have felt a sense of admiration and awe for the way Ramanujan worked. He had no need of a university education. With all his problems, he went on doing mathematics, writing page after page of equations in his notebooks. What was the urge that made him do that? It was not money. He was in poor health. He had no real patrons until he proved himself, and even those patrons probably learnt more from him than he from them. How could he do so much in such a short life, amidst suffering and sacrifice?

One cannot help but marvel at the way Michael Faraday carried out experiments. Faraday, who hardly had any formal education, was a genius propelled by an urge to explore. He was painstaking and dedicated, and a storehouse of intellectual energy. He just would not stop. I have been baffled by the artistic productivity of Michaelangelo who took many years to produce his masterpieces on the extensive ceiling of the Sistine Chapel at the Vatican. He painted alone, without assistance, day and night. He painted for hours on end, often in the dark with a candle on his forehead and another on his shoulder, with his head bent backwards to face the ceiling. Upon the completion of this monumental work, he was temporarily deformed and could not bend his neck. He could not walk because his toe nails had overgrown in the shoes which he had not removed in years.

I have found science to be a demanding master, not easy to satisfy. Only those who completely submit to it seem to reap the benefits in terms of intellectual excitement and satisfaction. Those who treat science as a pastime or a hobby may get some rewards, but nothing else. For a real scientist, all days are working days. For someone absorbed in his work, there are no working hours. Wherever one is, and whatever one is doing, one is always under the effect of the undercurrents of one's scientific pursuits, consciously or unconsciously. When such a thing happens, the need for an external stimulus to pursue science disappears. It is only then that anxiety about recognition and rewards disappears. This mental state is necessary for philosophically well-adjusted living. The effort to attain this state is difficult and may demand considerable personal sacrifice, often in terms of social life; then, it is worth it.

There is no limit to excellence in scientific pursuit and this limitless world that scientists belong to makes life worth living and more challenging. I have always been thrilled by the way research areas develop as one pursues ideas. 'Great oaks from little acorns grow.' As Herbert Brown has written, 'What starts off as a mere grain of pollen develops into an acorn. The acorn then grows into an oak tree. The oak tree develops into a forest. We then begin to see the outlines of a whole new continent.' There are undoubtedly many such continents lying undiscovered around us. Much of the life of scientists is spent in search of the grain of pollen, or working in a forest. Happy are those who witness the growth of a pollen grain into a continent. This can happen by chance, but chance only favours the prepared mind; it happens mainly due to persistent effort. I have been making all effort possible in the last six decades to seek happiness through scientific explorations, and in this process, science has become a way of life. I am indeed grateful for this blessing.

In closing

To my knowledge, no other large democratic nation in the world has faced the challenges that we do in India. We have to feed the poor and, at the same time, compete with the advanced countries in science and technology. This clearly demands that our base in science becomes sufficiently strong so as to meet the challenges. We have to bring in scientific awareness among all citizens, educated or uneducated, rich or poor. We have to ensure that science and education provide the foundation for the nation's progress and development.

'I arrived again at science, when I realized that science was not only a pleasant diversion and abstraction, but was of the very texture of life, without which our modern world would vanish away. Politics led me to economics and this led me inevitably to science and the scientific approach to all our problems and to life itself'—Pandit Jawaharlal Nehru (1938)

'Subtle is the Lord, but malicious he is not.'

'Nature hides her secret because of her essential loftiness, but not by means of ruse.'

'It is difficult even to attach a precise meaning to the term "scientific truth". The meaning of the word "truth" varies according to whether we deal with a fact of experience, a mathematical proposition or scientific theory'—Albert Einstein

'Even if I could be Shakespeare, I think that I should still choose to be Faraday'—Aldous Huxley (From Nugel, Sexton, Meckier)

Acknowledgement: In writing this chapter, I have made extensive use of my article entitled '*Doing Science as a Way of Life*', published in *Current Science*, 64 (1993), pp. 288–293.

3. EARLY YEARS: DAYS OF INNOCENCE AND HOPE

I was born on 30 June 1934, in Bangalore, at my maternal grandparents' house, close to Bugle Rock in Basavangudi. The house was in the vicinity of many temples, the bull temple being the most famous. A popular Ganesha temple was also nearby. My mother, Nagamma, apparently used to go to the Hanuman temple in front of the house every day. My father, H. Nagesa Rao, was employed in the education department of the Mysore government and worked for many years in high schools and teacher training institutes, and as an inspector of schools. My father was the

Rao's mother, Nagamma, at her daily prayers

Rao's father, Nagesa Rao

first person to go to college in his family and my paternal grandfather lived on the meagre income from the land he owned near Chintamani. My maternal grandfather was a teacher of Kannada and Sanskrit.

I remember little of my early childhood, except that my parents paid a good deal of attention to me. Clear memories of my childhood start from somewhere around 1940, when I joined the first year of a government middle school at the age of six. Till then, I had studied at home—under the tutelage of my mother, who taught me arithmetic and other subjects. In addition, she used to teach me Indian history and mythology. She was a fantastic teacher.

The medium of teaching in the middle school was Kannada. I was the youngest pupil in my class, but I was quite good at my studies and liked my teachers. I was proficient in English, thanks to my father, who often spoke to me in English and made me write letters and short essays in English. My years in middle school were uneventful. My companions came from ordinary families and I was one of the few students with a highly educated father, who had master's degrees in history, education, etc. I remember my friends asking me, 'Will you get a BA when you grow up?'

My father and mother were a study in contrast. My mother prayed for hours and was a great believer in the Madhwa philosophy. She was generous to a fault and had no interest in material possessions. My father did not conform to traditional practices and was against superstitions. He championed the cause of education for women and family planning. He did not allow horoscopes in our house.

There were very few schools and colleges those days, even in a city like Bangalore. There were only three government high schools and half a dozen private ones. It is possible that the British only allowed a minimum number of schools and colleges to come up in India—just enough to help administer the empire. An event I remember from my middle school years was the passing away of Rabindranath Tagore in 1941. My mother told me about the greatness of Tagore and his literary accomplishments. The year 1942 was also important in the history of India. The freedom movement had reached its peak. I used to see processions and often heard of people being arrested and crowds being lathi charged by the police.

Mysore state was a quiet and happy realm to live in. Bangalore was the capital of the princely state and was truly a garden city. The maharaja lived in Mysore. The annual Dasara festival and the procession, with the maharaja sitting on an elephant, were enjoyable events that people looked forward to. The weather was pleasant and there were no fans in houses in Bangalore. Most people walked everywhere and only a few buses would ply in the city. I walked to my school, which was about two kilometres from home. One day, when I was playing with my friends, my father called me to tell me that I had passed the lower secondary (LS) examination in 'first class'. The LS examination was a public examination given at the end of class seven. I received a few presents from my grandmother and great-grand-aunt. My father treated me to an ice cream.

After passing out of middle school in 1944, I joined a high school, which had a Kannada-medium option. My father was emphatic that his children should study in their mother tongue at school. High school education was for three years, at the end of which one had to take a public examination to receive the secondary school leaving certificate (SSLC). The SSLC was the qualifying examination for everything one did later. Depending on the marks obtained in the final examinations, one was declared as 'eligible for public service or eligible for college and public service' (EPS or ECPS). EPS was enough to become a clerk or a school teacher. One had the option of taking science or humanities for special study in high schools. I opted for science, which involved more courses in physics, chemistry, biology and mathematics. My three years of high school were spent in three different schools since my father was transferred to different places. I had good teachers, especially in chemistry, and some of them showed experimental demonstrations in the class. The very first experiment I saw demonstrated in the class was in my first year of high school, in 1944. He prepared hydrogen and burnt sulphur in the laboratory. He gave me a small amount of sulphur to burn over a home-made spirit lamp in my house. He made me build the spirit lamp with a used ink bottle. A year later, one Saturday morning, I helped my chemistry teacher prepare chlorine. My teachers in high school were excellent and I did not have such inspiring teachers when I went to college. It was during this period that I met some students from the Indian Institute of Science. I used to talk to them about what they were doing, and still remember one of them telling me about his research in chemistry. He took me to the institute's campus. I walked around with him, but did not understand what exactly was happening there.

In 1945, I had the wonderful opportunity of visiting the Indian Institute of Science when I was studying in Acharya Pathashala in Bangalore. Professor C.V. Raman visited our school and unveiled a photograph of Madame Marie Curie. His lecture was electrifying. I can still see Raman, holding the lapels of his coat, talking spiritedly about science and Madame Curie. After the lecture, he asked our teacher to bring two or three good students to his laboratory at the Indian Institute of Science. I was

Prof. C.V. Raman

one of the students selected to meet him. Professor Raman went around the campus with us and spent nearly half an hour with us in his laboratory. I did not understand a word of what he said about physics, but I was completely enamoured with what I saw. I thought, 'My god, how wonderful it would be to be like C.V. Raman!' It was probably then that the seeds of my becoming a scientist were sown.

During my high school days, I participated in literary activities. I used to write little stories and poems in Kannada, and took part in debates. My mother and I would read novels written in Kannada by great writers.

Holidays meant spending time with my maternal grandparents. I enjoyed the wonderful company of my grandmother, who was a great conversationalist and was always full of stories. She was a happy person and a great cook. She was extremely generous and had no control over the number of guests and freeloaders who came home to eat or stay temporarily.

I visited several places in the interiors Mysore state with my father when he went to inspect schools. This enabled me to see the forests and plantations of the Malnad region. An extraordinary aspect of my parents was that they consulted me on all matters, even when I was very young. They treated me like an adult and trusted my judgment. My father influenced me to develop the reading habit. He would encourage me to read general literature as well as the classics. He made sure I read books like *The Three Musketeers, A Tale of Two Cities, The Hunchback of Notre Dame, The Count of Monte Cristo,* and so on. This helped me communicate in English effectively. My mother constantly gave me one piece of advice: 'You must only get benefits bestowed on you by Saraswati (the goddess of knowledge), and not Lakshmi (the goddess of wealth). Everything in life must come through knowledge.'

The freedom struggle was in full swing during my high school days. I passed my SSLC examination in first class in 1947, when I was thirteen years old. I then joined a government intermediate college with physics, chemistry and mathematics as special subjects. The medium of instruction was English, but I had no difficulty with the courses. Unfortunately, those who opted for the combination of physics, chemistry and mathematics had no opportunity to study biology. I often felt bad that I wasn't studying any biology.

A few weeks after I joined college, India gained independence. The date was 15 August 1947. I can never forget the excitement of the first Independence Day the country celebrated. I still remember the speeches of Pandit Nehru, our first Prime Minister, and others. There was a mini-freedom movement in Mysore state soon after, since our maharaja did not agree to join the Republic of India straight away. I participated in this movement, which lasted a few months. I even used to wear a Gandhi cap those days. One cannot forget how food and cloth were rationed at that time. One would only get a limited amount of rice, kerosene and cloth. Those were difficult days, but they were also the days we all came together to pray for a great new India.

30 January 1948 was a terribly sad day. Gandhiji was assassinated by a mad man. I could not believe it. I was moved by the speeches made by Nehru and others after this monumental national tragedy. I kept trying to remember how Gandhiji looked. I had seen him once when I was eleven years old.

Pandit Jawaharlal Nehru

After completing my intermediate examination, I decided to do a BSc degree at Central College in Bangalore. Many of

Rao as a young college student (1947)

my friends went to engineering colleges, but I was not tempted to join them. Central College was the only college in Bangalore from where one could get a science degree at that time. There was only one university in the state, the University of Mysore, of which Central College constituted the science centre. Classes at Central College started in the morning around 11 a.m. and most students would walk to college after having an early lunch. Money was hard to come by and most of my classmates did not have the money to even have a cup of coffee during the afternoon break. There was still food rationing and restaurants did not serve popular items like dosa and idli, much to the horror of Bangaloreans.

Central College, Bangalore

There were a few lectures in the college each day, but I do not remember anything special about them. Of all the subjects, I had better lecturers in physics. I participated in chemistry seminars and gave one on polymers. I also took an interest in the Sanskrit Association of the college and in the debating society. The period between 1949–51 were the early years of independent India, and we were highly influenced by our national leaders. During my days in Central College, most evenings were spent with friends— talking about India, freedom and our future. I used to attend evening lectures at the Gokhale Institute of Public Affairs and the Institute of World Culture regularly. It was common to see leading literary personalities roaming around in the neighbourhood.

It was difficult to do well in college examinations those days. We were not given high marks, however good the answers were. For a perfect answer, one would get six or seven out of ten, rather than ten out of ten. I do not understand why it was so. The situation was worse in language examinations. The fear of the final examinations used to be quite intense as they were only held annually, and not at the end of each term or semester. Then there were the 'friends' who frightened one with questions all the time or kept talking about the most likely questions that would appear in the examinations. One had to be deaf or immune to such conversations.

My father used to warn me, 'I don't think that you will get a first class in your BSc. Hardly anyone gets a first class in BSc, but you must try.' When the results of my BSc examinations were announced in June 1951, I found I had obtained a first class, one of three students with a first class that year. I was seventeen years old then.

During my BSc course, I had a teacher in chemistry with a master's degree from Banaras Hindu University (BHU). He showed me a research paper based on his thesis at BHU. He told me that the MSc course at BHU was based partly on research

Banaras Hindu University, Varanasi

and that one could even publish a paper or two based on one's research. This impressed me. One could obtain a master's degree entirely through research from Bombay University. The diploma programmes in engineering at the Indian Institute of Science did not interest me. I thought that a combination of courses and a thesis would be ideal for me. I decided to go to BHU for my master's degree. I sent a letter, along with my BSc results, to the department head (Professor S.S. Joshi). My teacher, the BHU alumnus, wrote a letter of recommendation for me. In about a week's time, I received a telegram stating that I was admitted to the MSc class in chemistry at Banaras Hindu University. My father approved of my decision, though it meant sending me Rs 75 every month for expenses. In retrospect, I feel that my parents were fantastic in supporting me and trusting my judgement. They did not ask me why I was going for an MSc in Banaras, instead of doing something else in Bangalore itself. I admire them for the support they provided me.

In June 1951, I went to Banaras. I reached on a hot afternoon, after a long train journey, lasting over two days, in an unreserved third-class compartment. I had not been to north India before and had not seen a cycle rickshaw ever. I had to take a cycle rickshaw from the railway station to the university campus. After a couple of days, I could settle down in a hostel. The food was different. I was introduced to samosas, dal, rotis, palak paneer, rasdhar and gulab jamun.

In the first week of the term, I was asked to meet the head of the department, Professor Joshi, who was a DSc (London) and a student of Professor Donnan. He asked me what I was going to specialize in. I told him that I wanted to be a physical chemist. He asked me to start work on my MSc thesis right away. 'You work with Dr Marathe, who has just finished his PhD. He will help you to get started,' he said. Dr Marathe was a physicist and helped me set up some experiments on the effect of light on electrical discharges (what was called the 'Joshi Effect'). Within three weeks of joining BHU, I had started research for my MSc degree. Research was possible only at night, since there was only DC electricity and one had to convert it to AC electricity using a converter. This could be done only at night. I would go to the department around ten in the morning and spent the day doing laboratory work that involved the analysis of inorganic materials, organic preparations, as well as

The chemistry department at BHU

organic qualitative and quantitative analysis. There were also experiments in physical chemistry that I carried out. It was quite tiring, especially on days that were particularly hot, and there was hardly any time to have tea during the day. Once in a while, I would sneak out of the laboratory with a couple of friends to have a cup of sweet tea sold on the roadside next to the physics building. I was somewhat nervous in the first few weeks because of the tall claims made by my fellow students regarding their undergraduate training. Gradually, I managed to find an equilibrium. I got busy and did not worry about other matters. The early experience at BHU was enjoyable, as I had not worked in a laboratory before, spending long hours doing experiments. During my undergraduate days, laboratory work would last for an hour or two. Real involvement with chemical experiments at BHU and working at night in the research laboratory added to my experience. Unfortunately, there were only a few lectures during the term.

I started getting some results on electrodeless discharge in hydrogen after a few weeks. After six months of work, Dr Marathe told me that I had done enough work for the MSc thesis. I, however, went on doing something or the other on my own. When I was doing so, there was a visitor from Holkar College, Indore—a Dr Saxena, who was visiting the university to work with Prof. Joshi. Prof. Joshi asked me to

Rao at the Banaras Hindu University (aged eighteen)

work with him and help him do some experiments. I worked on silent electrical discharge in iodine vapour and made some measurements with him. Dr Saxena said that he would like to publish the results in a journal. He published a paper with me in the *Agra University Journal of Research* in the 1954 volume. This was my first research paper. After the first year of study, I had to take up the final examinations in the minor subjects (organic and inorganic chemistry). A month later, when I was in Bangalore for the summer holidays, I was pleasantly surprised to learn that I had stood second in class in the aggregate marks in the first-year MSc examinations.

During the second year of MSc, I was to study only physical chemistry. While my classmates started out on their MSc theses, I had already got mine ready. I was, however, allowed to do research on my own. I took the MSc final examination and submitted my thesis to the university in May 1953.

I had to give seminars during my MSc studies. I gave one on cosmic rays. I thought it was an exciting subject and was the talk of the town in physics. However, very few people seemed to know about this in chemistry. Another seminar of mine was on atomic hydrogen. I got used to the jokes that some of my classmates made about me. They used to call me 'professor', since I often talked about scientists like Linus Pauling, Peter Debye and so on. I was extremely excited about Pauling's work on

Linus Pauling

the alpha-helix. I had been truly impressed by his famous book *The Nature of the Chemical Bond* which I read after borrowing it from one of my professors. While people made fun of me, I was utterly serious about what I was saying about Linus Pauling and the kind of chemistry that he had done. My only hope was that one day I would do chemistry of the kind I read in his book.

Early in June 1953, I learnt that I had passed my MSc with a first class, again with the second rank in class. All in all, my Banaras experience was fantastic. BHU was truly a national university. It was there that I decided to become a scientist (and a chemist). It was in Banaras that I saw and heard great scientists (chemists) from abroad, like Lord Todd. It was there that I got properly exposed to classical music. Even though I came from a family where several relatives knew classical music, I myself could not learn music. Thanks to the music college at BHU, headed by Pandit Omkarnath Thakur, I could attend many fine concerts. That is where I first heard great musicians like Vishnu Digambar Paluskar. Living in the BHU hostels with students from all over India was an enjoyable experience, as were my occasional visits to the city. I first tasted the sweet Banarasi pan at BHU. I also got used to taking cold water showers through the year.

In terms of academics, although there were hardly any lectures given by the faculty members to postgraduate students, and the teaching laboratories were poor in instrumentation, the overall atmosphere was positive and encouraged one to pursue research. The laboratories in the science departments were lit up at night with people doing research. Prof. Joshi hardly looked at my research work, but he was an enthusiast and always talked of research. Staying in Banaras was also a spiritual experience for me. Banaras has something more to it than other cities. It is an eternal city, a city of light. Banaras had made an impact on me beyond measure.

I was asked whether I would like to take up a temporary lectureship at BHU, but I was not interested. I wanted to do research and was looking for a possible place to work on a PhD degree. After an MSc in physical chemistry, my hope was that I would be able to do research in a subject of current interest at that time. This meant working in a subject like molecular structure, of which Pauling was the master. It was not clear to me how I could do this in India. When I met the professor of physical chemistry at the Indian Institute of Science (IISc), he told me that he did not think that the physical chemistry being pursued there was of current interest. He advised me not to join the department at IISc. I was surprised—here was a professor of physical chemistry telling me not to join his own department because it was not doing the kind of physical chemistry that was at the frontier globally. He asked me, 'Why don't you join the Indian Institute of Technology (IIT) at Kharagpur for a short while?'

IIT Kharagpur was just coming up under the directorship of Sir J.C. Ghosh, who had left IISc to start the first IIT at Kharagpur. A number of faculty members from IISc had joined the new IIT. Prof. Ghosh had a good reputation as a physical chemist. I was interviewed in Bangalore and selected to work as a research scholar in the director's laboratory at IIT Kharagpur. The institute was located

Sir J.C. Ghosh

(From Wikimedia Commons)

in an old building, near where the construction of a new campus was just getting started. Kharagpur had a large railway colony and had nothing much to offer. Calcutta (Kolkata) was around two or three hours away by train. Dr M.V.C. Sastri was my immediate supervisor and I worked on the 'adsorption of gases on catalyst surfaces'. I built equipment for this purpose and did some experiments, and, in the meantime, also started writing to chemistry departments in the US where I could do my PhD. I was in a hurry to learn chemistry of the kind that would enable me to understand the articles published in major journals. I also published a note based on my BHU thesis in a British journal. An important gain of having lived in Kharagpur was that I got to learn about Bengal and Bengali culture. I was amazed by the renaissance period of Bengal. My love for Bengal, which was kindled then, continues till today.

The Indian Science Congress had its annual session in Hyderabad, in January 1954. I had sent the abstract of a paper that I wanted present there. As Professor Linus Pauling was to be the main speaker at the congress, I thought that I should at least see the great man and went to Hyderabad in a train, in an unreserved third-class seat, at my own expense. I was terribly disappointed when Professor Pauling did not show up. He had been denied a passport and could not leave the US. He was then considered a 'commie' and a security risk because of his opposition to atmospheric nuclear testing. By that time, I had written to him about the possibility of working on a PhD with him on molecular structure. He wrote back stating that he was no longer doing work on molecular structure and suggested that I could work with one of his former students elsewhere. It was in this connection that I learnt about Purdue University, where Professor Livingston worked on electron diffraction of gases to determine structures of molecules. Late in 1954, I received an offer of a research assistantship ($150 per month) from Purdue along with admission to the PhD program in chemistry. I thought it was a good idea to go to Purdue, instead of MIT and Penn State, both of which had also offered me assistantships.

I told my father about my plans to go to the United States for my PhD. The only thing he said was, 'Don't you think that you should have told me about this a few months earlier? How do I get money to support your travel and so on, at such short notice?' He started working on ways to help me go to the US. I had to get a passport and that was not easy. I applied for one from Kharagpur. I had to travel to offices in Midnapur and get a variety of certificates and guarantees. My father's financial guarantee was not enough. Fortunately, a faculty member at IIT Kharagpur gave me a guarantee, to obtain the passport. I could get the US visa from the consulate in Calcutta after a medical examination and taking the loyalty oath. I was all set to go to the US.

I had to go by boat, first to London and then on to New York. I got a berth in SS *Corfu* from Bombay to London and then by SS *Queen Mary* from Southampton to New York. It was a seventeen-day trip from Bombay to London.

Rao with his parents before leaving for the US

My father had come with a few friends to see me off. I was deeply touched to see my father crying as the boat left the shores of India. I was the only student on the boat. I ate eggs for the first time, at breakfast, and saw people dancing in the evenings. I was in a completely different world though and was only thinking of my PhD studies in the United States, and of the orientation exams that I would have to take when I reached there. Through the journey, I tried to brush up my knowledge of physical chemistry, organic chemistry and so on. When I reached London, I was helped by an acquaintance to settle down in a small hotel. London was still recovering from the damages of the Second World War. I spent three nights there, before taking the boat to New York.

I reached New York after five days. I took a taxi to the YMCA, as advised by a friend. YMCA had a public bathing place where men took showers. I was not used to being in the nude in public. I decided to take a shower very early in the morning, when no one would be there. I stayed there for two days and spent one day taking in the sights of the city. I then took a Greyhound bus all the way from New York to Lafayette, Indiana, since it was the cheapest way to get there. At one of the stops, at a food counter, I saw a nice round thing with a hole in the centre, and thought it was a vada. I bought one only to find it was sweet. I had not seen doughnuts before. In fact, I had not seen or heard of many things before in my life. More importantly, I had not experienced the rigour of a good education in chemistry. I was anxious and worried. I wanted to do well in the university. I badly wanted to become a good scientist.

After arriving in the US, I became concerned about my finances. My father had given me just $150 for the first month's expenses, exactly how much the research assistantship would provide a month later. I had to be careful. I reached Lafayette, worrying whether I was up to the standards expected of a PhD student. My worry was mainly because of the difficulty I had in understanding the language and the contents of the papers published in mainstream journals. The day after I reached Lafayette, I went early in the morning, at 8 a.m., to the chemistry department, wearing the new suit that my father had got stitched for me, complete with a necktie. I was the only one dressed this way. The other students were dressed casually. We were to take up orientation exams in the four branches of chemistry.

The day after the exams, the convener told me that I had done fairly well in organic, inorganic and analytical chemistry. The problem was with physical chemistry, which was supposed to be my major subject. They had given only problems to solve in the examination and I was not used to solving problems quickly. We wrote essays in India and never solved problems. I was using logarithmic tables to do numerical calculations, when everybody else was using a slide rule, which I had never seen before. I had just managed to pass the examination. I was advised to take several courses and opted to take one in standard physical chemistry involving problem solving, in addition to advanced courses in analytical chemistry and inorganic chemistry. I also took a course in

X-ray crystallography in the physics department. I had mentioned to the department that I would like to take up the chemical physics programme, majoring in physical chemistry with physics as the minor subject. I knew this was going to require hard work since I would have to take courses in physics as well as in advanced physical chemistry, and pass the preliminary examinations in both physical chemistry and physics. I was warned by some senior students that such a programme may require a good six years for the PhD degree. Even so, I decided to take up these subjects.

There were only a few Indian students at Purdue when I went there—just ten of us in all. One was doing a PhD in chemistry. Most others were doing undergraduate or masters' programmes in engineering. The situation changed quite quickly. Three years later, when I was leaving Purdue, there were nearly sixty to seventy Indian students. Lafayette had little to offer in terms of restaurants or entertainment. The programmes organized by the university in the music hall were the only occasions for cultural entertainment.

Course work at Purdue was demanding. In the very first semester of my stay, in addition to the chemistry and physics courses, I audited a mathematics course to improve my background in the subject. Each course had three examinations during the semester and one final examination. With so many courses, I had to keep preparing for one examination after the other. In the meantime, I also had to do some research as well.

The Chemistry department at Purdue University

Although I had opted for a major in physical chemistry for my PhD, and had decided to work on gas-phase electron diffraction to investigate structures of molecules (with Professor R.L. Livingston), my research assistantship was with a professor of organic chemistry, Professor Eugene Lieber. Professor Lieber wanted a physical chemist to study kinetics of reactions and spectroscopy of some of the molecules made in his laboratory. I started working in his laboratory a week after I joined Purdue. I could only work at night and during the weekends. To keep abreast of the courses and exams, and also contribute to research, was a formidable task. However, along with Dr Chao, who was working with Professor Lieber as a postdoctoral fellow, I could accomplish something in my first few weeks. This involved a new synthesis of alkyl azides. Professor Lieber immediately sent a paper to the *Journal of Organic Chemistry*, which was accepted. In the meantime, a number of triazoles were made in the laboratory and I studied kinetics and equilibria of certain reactions involving them, and examined the effect of substituents on the reactions. Professor Lieber published these results in two papers in the *Journal of American Chemical Society*. The first problem involving spectroscopy that I worked on related to the infrared spectra of organic azides. I could assign the symmetric and asymmetric stretching modes, as well as the bending modes of the azide group. This paper was published in *Analytical Chemistry* and has been cited by a number of chemists since then.

I was very busy the first semester, with hardly any time for pleasure. Slowly, I got used to the hectic routine. I realized that hard work was something one had to get used to. Hard work and continuous effort depend on the commitment of the individual. These are the two qualities I tried to imbibe in the first semester at Purdue. At the end of it, I found I had done well in all the courses, including the one in physics, and had scored 'A' grades. I did not do as well in the physical chemistry course, the subject of my specialty. This was again due to the fact I could not solve all the problems in a given period of time, because of the slow rate at which I wrote my answers. I used to write answers in great detail, explaining why and how I solved the problem. What was required was to solve problems by just writing a few steps and give the answer. I decided to improve my problem solving abilities and within the next couple of months, I mastered the art of solving problems at a faster rate. The following semester, when I took courses in advanced thermodynamics and so on, I did very well. I remember the advanced course on instrumental methods of analysis. There would be ten to fifteen questions in an examination, which had to be answered in an hour. I did fairly well from the beginning and scored 100 out of 100 in the last two examinations. I enjoyed the course on advanced organic chemistry taught by Professor Bob Benkeser and the advanced inorganic chemistry course taught by Professor Herbert C. Brown. The organic chemistry course was so good that it made me feel like I had mastered organic reactions and synthesis. The knowledge of organic reactions gained then still continues to be useful to me in research.

By September 1955, I had published a few papers with Professor Lieber and had not yet visited my major professor (Professor Livingston), with whom I was to do electron diffraction studies. Professor Lieber took me to the fall meeting of the American Chemical Society in Minneapolis. I had never been to a professional meeting of that kind before. I saw many great chemists in the meeting and one person I particularly remember seeing there was Peter Debye.

During the first year of my stay at Purdue, I learnt many new concepts and techniques in chemistry. For example, I had learnt how to quantify the effect

(From www.nobelprize.org)

Prof. Herbert C. Brown
(Nobel laureate in chemistry)

of substituents on chemical reactivity and decided to use this knowledge to understand spectroscopy of organic molecules, in particular, electronic and vibrational spectra of benzene derivatives. I carried out simple correlations during my spare time and published short papers on my own on the topic. Professor H.C. Brown encouraged me in this effort and took an interest in my work, even though I was not his research student. The number of my independent publications started increasing with time. I collaborated with fellow students and faculty members to solve problems that were mentioned in the classroom or in seminars.

At the end of a year, I had to take qualifying examinations in all four branches of chemistry. This was a requirement of the chemistry department, independent of what one specialized in later. As a chemical physics student, I had to take a qualifying examination in physics as well. I took the examinations in chemistry and passed, much to my delight. That was a big relief. I did not expect to go through the qualifying examinations so smoothly. Just after the qualifying examination, some of my friends told me they were are going to New York for a short visit. They asked me whether I would like to join them and watch the famous musical *My Fair Lady*. I went for three days and stayed with a friend's family.

The research assistantship with Professor Lieber enabled me to learn a great deal of chemistry—of the kind I would not have normally come across as a physical chemist. It also gave me several research publications. Unfortunately, Professor Lieber decided to leave Purdue and go to Chicago at the end of my first year. My research assistantship with him came to an end and I had to teach for a semester in the department. I was a teaching assistant in physical chemistry, which also involved taking care of the laboratory course. Professor Lieber wanted me to continue to be associated with his research and invited me to visit him over the weekends at least once a month. I often stayed overnight in his house during these visits. I continued to collaborate with him and his students on spectroscopy, and published a few more papers with him. Many Indian students worked with Professor Lieber and one of whom was J. Ramachandran (Ram). He later studied biochemistry at Berkeley and became director of research in Genntec, in California, and President of Astra India. Ram's first few papers were with me.

When I met my research supervisor, Professor Livingston, he mentioned that we had to build a new electron diffraction instrument as the old one was not in good shape. He wanted to use a rotating sector for recording the diffraction patterns. We were to set up the instrument in the new chemistry building (Wetherill Laboratory). I was a bit worried about the time it would take to build the apparatus. Fortunately, one of the students senior to me was good at building things. I participated in some aspects of the fabrication. For example, I made rotating sectors that could be fixed above the photographic plates. After six months, we had a new instrument ready for recording diffraction data.

I completed my preliminary examinations in physical chemistry and physics, and also the examinations in technical German and French in 1957. The prelims included both written and oral examinations. Questions were asked based on an original proposition submitted by the student. The original proposition was to be a research problem the student would work on as an independent scientist. I had a disagreement with a faculty member regarding a question in physical chemistry, but eventually we resolved it and became friends. During this entire period, I continued to do something or the other on my own, and published papers on infrared and electronic spectroscopy of molecules. I published some of them in India too, so that I could be connected to India in some way.

As I started getting a bit concerned about the time I would require to complete my PhD thesis, one fine morning,

I received a letter from the head of the department saying that the department had decided to award me the Standard Oil Foundation Fellowship, given to the best PhD student in the department. The salary was still $150 per month, but I felt good about being recognized. By then, I had also been elected the graduate student representative in the department. I had a large group of friends and I was closest to Harry Rubinstein, Sandy Kern, Joules Kristoff, Ed Kearns, Cal Kobayashi and Qi Che. Amongst the Indians, A.K. Ramdas, who had come to Purdue as a postdoctoral fellow in physics after his PhD with Professor C.V. Raman, was a close good friend.

In gas-phase electron diffraction studies, I worked on the structure of two molecules of great interest to Linus Pauling—carbon suboxide and methyl azide. I could solve the structures of these molecules in a few months. The structures validated the adjacent charge rule of Pauling. Around that time, Pauling visited the chemistry department and saw my results on the two molecules. He was delighted and cited the results in the third edition of *The Nature of the Chemical Bond.* I also solved several other molecular structures. Professor Livingston was then a Dean and was busy in administration.

Prof. Livingston with the electron diffraction apparatus built by Rao and other students at Purdue (1956)

He depended on me for running the research group and had become a friend. He used to take me out for lunch regularly or would invite me home for dinner. I worked with every student in his group and most papers had my name on it. I ended up publishing around ten papers on studies of molecular structure by electron diffraction. One of the problems I studied related to the structural predictions of Pauling on chloroethylenes. Another important problem was on the role of hyperconjugation in organic molecules, if any. In September 1957, I thought that I had done enough to get a PhD and Professor Livingston was glad to let me submit my thesis in January 1958, just around three years after I joined Purdue. Most students solved one structure for their PhD thesis, but I had worked on six structures. This was wonderful because some of the students who had joined the chemical physics program along with me had barely started their thesis work.

It was customary for students to wear a proper suit for the final PhD examination. I did not have a good suit and had to borrow $25 from a friend to buy one. When I went into the examination room, the four examiners greeted me with smiles. What do you plan to do next, they asked. Before I replied, they said they were delighted to pass me in the examination. It was over in less than ten minutes. Some of my friends thought I had flunked when I came out of the room so quickly. I was fortunate to have finished my PhD work early and, at the same time, to have gained experience in several areas of chemistry. I was less than twenty-four years old when I completed my PhD degree. My parents were overjoyed by my becoming a doctor of philosophy. For me, it was the first step in a long journey. By the time I finished my PhD, I had around twenty research papers, of which only five belonged to my PhD thesis.

I must mention about a paper of mine from the early days. A year after I started my PhD work at Purdue, I decided to write a note on some of the independent work that I had done during my MSc studies at BHU, in answer to a paper that had just appeared in a reputed journal. I sent a note to the journal *Science*. Amazingly, it was accepted and the article was published in the journal in 1956.

As I was completing my PhD work, I could work on a problem in solid state chemistry, thanks to my friend Al Czanderna, who was a student of George Honig. He had prepared spectroscopically pure TiO_2 and expected the structure to be that of rutile. I told him that I would determine the structure for him using X-ray diffraction. I found that the TiO_2 sample had the anatase structure. That was a surprise and we decided to study the transformation of pure anatase to rutile as a function of temperature and other parameters. This resulted in one or two nice papers. I became interested in the subject of phase transitions because of this and continued to work in this general area in my later years. Throughout my studies at Purdue, the physics department, in particular Professor Yearian, was kind to me. He allowed me the free use of his X-ray crystallography laboratory. This was a great boon to me. Professor Yearian often dropped in to the laboratory late in the evening when I would be working. He would ask 'How about a movie?' and take me to a movie theatre in town, generally for a double feature show.

By the time I finished my PhD, I felt I had carried out research on some interesting problems, with reasonable accomplishment. I needed some experience as a postdoctoral fellow in a good place, before starting an independent research career. I was a young man in a hurry. I wanted to finish the postdoctoral stint as early as possible. While I was waiting to decide on the postdoctoral fellowship, I was supported by a National Science Foundation fellowship. In addition, I was also teaching physical chemistry to undergraduates. For the first time, I was financially stable. Otherwise, I would be short of money by the last week of every month during my PhD days. I could not afford to eat in restaurants or even in the union cafeteria. I would eat a toast for breakfast in my apartment and home-made sandwiches for lunch. I had coffee in the students' union two or three times a day, where a cup cost 10 cents. The laboratory kept me busy and cheerful. By the month end, I would often borrow $10 from the office of the Dean of students and return it a week later. My working hours those days were, 8 a.m. to 1 a.m. I often took my morning shower in the chemistry building.

The chairman of the Indian Public Service Commission visited the United States and some of us met him. He stated quite rudely that he had no respect for American degrees. He said that even a doctorate from Oxford got Rs 300 or 400 per month as their starting salary. He did not see why American PhDs should be given any special consideration. I had to explain to him that American training for PhDs was far superior in many respects, but he was not convinced. He said that at that time 'even a DSc in India gets Rs 400 as their starting salary'. I did not know what DSc meant. I asked people and learnt that a DSc was a higher degree than a PhD and one could get the degree by submitting independent research papers to one's mother university in India. I immediately felt that I must somehow get a DSc degree if that was a requirement for getting a reasonable position in India. I found out that Mysore University accepted DSc thesis from those who had a BSc degree from there. Apart from the papers from my PhD thesis, I had a few papers of my own, and another dozen or so along with Professor Lieber, which could form a DSc thesis. I wrote an introduction, bound it with reprints of all the papers I had published and sent it off to Mysore University, along with the registration fees. Even now, when I think of it, I feel amazed at my courage then. After submitting the thesis, I forgot about it and became absorbed in finishing more work on electron diffraction and other areas at Purdue. I also started looking for possible postdoctoral positions.

I received an offer of research associateship from Professor Lipscomb. Unfortunately, he said that I would have to wait for a few months, since he was shifting from Minnesota to Harvard University. I was to work on proteins. I did not want to wait that long. I could have gone to University of Chicago, but, unfortunately, I was afraid to be in Chicago because the university campus was deemed unsafe at night. I decided to go to the University of

California at Berkeley, which had offered me a postdoctoral fellowship (with a monthly salary of $625) to work with Professor Kenneth Pitzer, a well-known physical chemist. Leaving Purdue was an emotional experience. Professor Livingston was in tears when I left Lafayette.

University of California, Berkeley

I liked going to California, which represented the more liberal region of the US. America in the early 1950s was still in the McCarthy era. The Rosenberg trial was going on. There was much conservatism in certain parts of the US. I occasionally experienced the effects of segregation. I was not served in a bar just outside the Purdue campus. Harry Belafonte, the famous entertainer, had also been refused entry to that same bar for being black. In Kentucky, I was asked in a bus station to go to the other side of the coffee counter, where blacks were served. In Salt Lake City, the hotel where I had a reservation denied having a room for me and asked me to go to the local YMCA instead. I just ignored such incidents.

Rao at Berkeley

Going to California was a refreshing experience. UC Berkeley was different from Purdue in many ways. When I went to Kenneth Pitzer to do postdoctoral work, I was not sure which areas I would work on. He suggested some problems on heat capacity of solids at liquid helium temperatures and on NMR spectroscopy. The heat capacity measurements were to be carried out in Professor Giaque's laboratory. I was also to work on NMR spectra of alcohols at different temperatures to understand hydrogen bonding. In

**Prof. K.S. Pitzer—
a great physical chemist**

addition, I studied the thermal properties of disordered metal oxides and investigated the electronic spectra of organic molecules. Professor Pitzer allowed me to publish some of the results independently. I had three or four independent papers, of which two were in *Nature*.

UC Berkeley was a fantastic place to be in. In Tuesday colloquia or Thursday seminars, the front row would be occupied by giants of chemistry. I still remember the day I gave my first physical chemistry seminar. The front row had Pitzer, Joel Henry Hildebrand, Glenn T. Seaborg, George C. Pimentel, Jura, Melvin Calvin and others. It was a frightening experience. One saw many great scientists in the corridors of Berkeley. Henry Eyring and other great physical chemists gave seminars regularly in the department. Those were the days when Linus Pauling was being harassed by the US government. I would see him frequently on TV having arguments with Edward Teller. Teller himself was a common sight in the Berkeley campus, as was J. Robert Oppenheimer. In that wonderful atmosphere, I thrived. I published a few of my own papers, thanks to Professor Pitzer who was very generous. I got to know many people, particularly Pimentel who was an outstanding teacher and scientist. Pimentel had obtained his PhD with Pitzer and was an assistant professor. My interest in hydrogen bonding got strengthened by my association with Pimentel. One could feel the ghost of G.N. Lewis everywhere in the chemistry department at Berkeley. Lewis started the Berkeley chemistry department in 1912 and nurtured it to grow into the best chemistry department in the world. I enjoyed

(From Edgar Fahs Smith Collection)

**Prof. G.N. Lewis—
one of his kind**

being in that department and became involved in newer areas of chemical science practiced there.

As this was going on, I was wondering what I should do next. I was starting to feel that I should start my independent career as early as possible. I was itching to work as an independent scientist with a group of students of my own. Such an independent career was possible in the US. I was approached by a few universities regarding faculty positions. One of the universities wrote that it would consider me for a promotion within two to three years after I joined, because of the large number of publications that I already had. As I was wondering about this, my correspondence with the Indian Institute of Science, Bangalore, fructified and I received a letter offering me a lectureship in the inorganic and physical chemistry department on a basic salary of Rs 500. This would give me a monthly income of around Rs 700 per month, including allowances. I also had an offer of a readership from Panjab University.

I spent many sleepless nights unable to decide whether to return to India or stay in the US. Everyone I knew in the US felt I would be making the biggest mistake of my life if I returned to India. Professor Livingston, Professor Pitzer and others felt I was best suited to the American academic system, since I worked so well with others and accomplished things at a fast pace (the 'American way'). I had to think about India and my parents, particularly my mother. I then received a letter from my father asking me whether I would consider coming back, since I had already been away for quite some time. I had not seen or talked to my parents after coming to the US. Visiting them was out of question because of visa problems. Phones did not work those days and the only communication I had with my parents was through weekly letters. I took the decision to join the Indian Institute of Science. This satisfied my desire to return to India and be with my parents. I thought that IISc was the best possible place for scientific work in India. This was important as I wanted to carry out my research in a place as good as the best anywhere else. This was my hope and prayer, but I was not sure whether it would be possible. As the days to leave Berkeley came closer, I was in a rush to finish several experimental projects. The day before I was to leave, I was making some samples for NMR studies. Professor Pitzer dropped in to the laboratory and said, 'Ram, we will really miss you.' He then took me home for dinner.

I decided to leave the US in September 1959, visiting Purdue University for a few weeks on my way back. I was anxious to join the Indian Institute of Science as early as possible. I believed in Faraday's famous statement that in science, we 'work, finish and publish'. H.C. Brown used to say that if it was worth doing, it must be worth publishing. I had a great urge to go back to India and start publishing good papers from India with my students. I wanted to do research on various problems of interest. In doing so, I would follow G.N. Lewis's famous definition, 'Physical chemistry is anything that is interesting.' In infrared spectroscopy, I wanted to examine matrix isolated species at low temperatures, band intensities and phase transitions, besides making band assignments of certain molecules based on normal vibration analysis. In electronic (UV-Visible) spectroscopy, I wanted to study excited states, solvent effects, molecular interactions and so on. If time permitted, I would work on some problems in solid state chemistry. I wanted to badly write a monograph on UV-Visible spectroscopy of organic molecules with molecular orbital notations. I was not sure what I would face in India, but was determined to start a new chapter in my professional life.

In closing

I left the US from Chicago by a TWA flight to London. After a short stay there (and a visit to Oxford to visit Dr Sutton to see his electron diffraction set up), I went to Paris and Geneva for a couple of days each, before taking the flight to Bombay (now Mumbai). I had a long wait at the airport for customs inspection, where the contents of my suitcases were thoroughly checked. By that time, the luggage I had sent from the US by sea had arrived and I arranged for its shipment to Bangalore. I arrived in Bangalore by an early morning flight. My parents, grandparents, uncles and aunts were all present in the airport. I was seeing them after a long, long time.

When one starts working in science, one thinks that there are a few steps to success, but as one goes on, it becomes clear that there is a limitless ladder of excellence. I wondered whether I was good enough to climb the ladder.

One thing is certain. One does not know what the future holds. As Niels Bohr stated, 'The peculiar feature of "prediction" is that one can never predict the future.'

Whatever the future may hold, one has to dream big.

'It is a tragedy if one does not realize one's dreams. It is a bigger tragedy if one does not dream at all'—Anon

4. DAYS OF STRUGGLE AND ANXIETY: PROVING ONESELF

The day after I returned to Bangalore from the US, I went to IISc. As I stepped into the beautiful campus, I realized that I was entering a new phase of my life. I was just over twenty-five years old, full of hope and enthusiasm. I met the head of the Department of Inorganic and Physical Chemistry, Professor M.R.A. Rao, and formally

Indian Institute of Science, Bangalore

joined the department as a lecturer. It was a small department with one professor, one or two assistant professors and three lecturers. This was the average size of most departments in the institute then. The department had recruited two students to work with me for their PhD degrees. In view of the limited space available, the department could not assign me any office space, but gave me a small laboratory in a corner of the building. There was a huge wooden structure in the middle of the room, a remnant from the older days. I was not to remove the structure. It left little space for laboratory work and almost no space for fabrication of equipment. I was expected to sit somewhere in the room and use it as my office. I was disappointed since I wanted at least a small office for myself. The department did not have sufficient funds and could give me around Rs 5000 for research.

I started scouting for facilities. There was no equipment worth mentioning in the department except for a manual UV-visible spectrophotometer. It took one or two hours to take measurements on a sample with the spectrophotometer. One had to measure the optical density point by point at each wavelength, and draw the spectrum by hand. There were no recorders in the department. I expected an infrared spectrometer to be available in the chemistry departments, but that was not the case. When I learnt that there was a single-beam infrared spectrometer in the physics department, I met the department head there. He was quite rude and told me, 'I don't see why chemists should use an infrared spectrometer.' He could not see any way of allowing me to use the spectrometer. I had several research publications on infrared spectroscopy, but that made no difference to him. I was in a terrible situation, not knowing what to do about research in spectroscopy.

As I was worrying about the available facilities, I noticed an old X-ray generator in the department with a tiny powder camera. One could get one or two lines in a diffraction pattern using the camera. I realized that I could explore problems solvable even with that X-ray equipment. The only spectroscopy that I could do was by using the manual UV-visible spectrometer. I initiated some work on solvent and substituent effects on the electronic transitions of organic molecules. One had to wait for ones turn to obtain spectra, but I tried to resolve the situation by working at nights, with special permission. I started publishing papers on electronic spectra, especially on $n\text{-}\pi^*$ and other transitions of non-bonded electrons in organic molecules. As far as infrared spectroscopy was concerned, I had to resign myself to the fact that I would not be able to work too much on it now. I thought that I would record spectra of some interesting molecules here and there and make band assignments. I went a few times to the National Chemical Laboratory, Pune, to record spectra and do library work.

Based on the spectra recorded elsewhere, one of my students and I could make assignments of the $C=S$ and $NH\text{--}C=S$ vibrations in molecules. This paper became a citation classic. My research based on infrared spectroscopy depended entirely on the facilities available elsewhere, that I could use through the courtesy of the concerned laboratories. I tried to explore whether I could use the facilities at the atomic research centre in Mumbai or at the Tata Institute of Fundamental Research. I did not get a positive reaction from either place.

I started doing some work on the phase transitions of the anatase and brookite forms of TiO_2 to the rutile form using the X-ray diffraction camera mentioned earlier. I published two papers on the phase transitions of TiO_2 in the Transactions of the Faraday Society over a period of two years. I then initiated work on rare earth oxides, which was a popular area of research at that time because of the problem of non-stoichiometry, as exemplified by Pr_6O_{11} and Tb_4O_7. I wanted to make pure stoichiometric PrO_2 and TbO_2 and study the various phases obtained by their decomposition. How could one make these dioxides? I had to find a way of doing this without the use of high pressures. I discovered that simple solvolitic disproportionation of Pr_6O_{11} and Tb_4O_7 by dilute acids would give the dioxides as solid residues. This meant that I had to just wash Pr_6O_{11} and Tb_4O_7 with dilute acids to get PrO_2 and TbO_2 as residues. This is an early example of 'chimie douce' and illustrates how research had to be based on what one could tackle with the marginal facilities available. If one wanted to do thermogravimetry, the way was to build a small balance using a quartz spring and record the data manually. That is exactly what we did. Solid state chemistry was not a popular subject in the 1950s and there were only a few practitioners of the subject in the world. I felt that I should pursue solid state chemistry more vigorously.

In the department, I organized seminars in the evenings, where students could talk on recent topics of significance. There was an enthusiastic response from the students, but, unfortunately, some of my colleagues became unhappy that their students participated in a seminar, which had no official sanction. I had difficulty in other academic activities as well. I wanted some changes in the course programme to reflect the modern developments in areas like quantum chemistry, molecular structure and spectroscopy. I wanted to give lectures on molecular structure and spectroscopy, as well as on chemical bonding, but the department did not encourage me to do so as there was an older member of the faculty who gave lectures on these topics, even though he had no experience or specific knowledge of the subjects. What was encouraging, however, was that many people in the country started asking me to give talks on spectroscopy and molecular structure.

In December 1959, I proposed to Indu, my wife. I called her from the departmental phone to get her consent, because of the lack of phones on the campus. Everyone in the department knew what was happening. Indu had a brilliant record as a student, with a BA (Hons) in English. Her father, B.S. Narayana Swamy, was a senior engineer in the Karnataka PWD. We got married in May 1960, in Bangalore, with lots of relatives in attendance. Indu came from the Bapu family, which was known to us. In fact, my maternal grandmother was related

**Indu and Rao on their
wedding day**

to the Bapu family. We lived in a spacious house in Malleswaram, quite close to IISc, paying Rs 100 as the monthly rent.

In 1960, I received a letter late from Dr Vikram Sarabhai, asking me to give a talk in the physics section of the Indian Science Congress at Roorkee, in January 1961, of which he was president. He wanted me to talk on spectroscopy in chemistry. I went to the science congress, and there I saw Pandit Nehru and several scientists, including S.N. Bose. I was lucky to have gone to Roorkee since I learnt there that a 40 MHz NMR spectrometer had been acquired by the Aligarh Muslim University, by Dr Venkateswarlu. I contacted him and collaborated with his group on two simple problems related to hydrogen bonding, and published papers jointly with them.

I was working under tremendous pressure, but was not exactly sure of what I wanted do in the long run. The professor of organic chemistry in the institute, Professor D.K. Banerjee, came to my rescue to some extent. He was a wonderful person and encouraged me in many ways. He gave me an excellent student of his department to work with me. He allowed me to use the small infracord that he had purchased for his department. We used to occasionally go eat lunch at the only Chinese restaurant in Bangalore on Brigade Road.

An important factor that helped me somehow maintain my sanity during those days was my close friendship with a few colleagues, especially S. Ramaseshan and Satish Dhawan. Ramaseshan was in the physics department and we used to commiserate every day about various matters in the institute. Satish Dhawan was a professor of aeronautical engineering and was a fine person with a high sense of morality. Social life was non-existent at the institute, but my wife and I met the Ramaseshans, Dhawans and Banerjees regularly. We enjoyed life, even though our financial position was precarious. My total salary was Rs 720 ($180 at the prevailing exchange rates) per month. After various deductions, I received an amount just enough to live a reasonable life. My father-in-law once asked me why I had not opted for a high-salaried industrial job, as the one that was being offered by ICI Ltd then. I had to tell him in jest that I could feed his daughter even with my salary and that I will not send her back to his house due to a lack of funds.

Late in 1959, I started writing a book on UV-visible spectroscopy of organic molecules. I had to go to the National Chemical Laboratory, Pune, for doing my library work since IISc did not subscribe to many of the journals I needed. There were disparaging comments made by some colleagues when I talked about the book. They thought it was a joke. The book was accepted for publication by Butterworths (London) and came out early in 1961. It was 170 pages long. When people saw printed copies of the book, they kept quiet. Many people, particularly Professor John Murrel of Sussex University, who was well-known in quantum chemistry and electronic spectroscopy, wrote to me, and later told me in person, that my book was the first to use molecular orbital notations to describe electronic spectra of molecules. The book has since been published in several languages and has had several editions. The year 1961 was also an important one at home. Our first child, Suchitra, was born.

One morning in 1961, I received a brown envelope by post. It was a letter from the University of Mysore. The university had awarded me a DSc degree based on the thesis that I had submitted in 1958, when I was in the US.

Soon after my book on UV-visible spectroscopy came out, I started to think about a book on infrared spectroscopy, which emphasized chemical aspects and discussed applications in organic, inorganic and biological chemistry in some detail. I started writing the book in the beginning of 1961 and completed it in June 1962. The book titled *Chemical Applications of Infrared Spectroscopy* (it was about 700 pages long) was accepted for publication by Academic Press, New York.

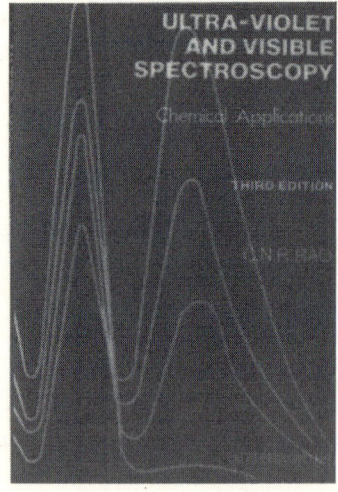

By 1961, I had taken a few more PhD students and initiated research on spectroscopy of hydrogen bonding and of charge transfer complexes, in addition to some aspects of solid state chemistry. Around this time, I was invited to give a talk at the international spectroscopy conference to be held in Maryland (USA), in June 1962. I decided to attend the conference and wanted to use the opportunity to spend a few months in the US. I was able to get a teaching position for the summer of 1962 at Purdue University, followed by a short-term research appointment in Chicago. I met Charles Coulson (the great theoretical chemist) at the conference and developed a close friendship with W.J. Orville-Thomas, who was then working at the University of Wales (Aberystwyth) in UK. The friendship continued for many years and we published a few papers together. Henryk Ratajczak from Poland was also close to us and we formed a team. Friendship with Henryk resulted in my trips to Poland and a few joint publications. Orville-Thomas is no more, though Henryk and I still correspond with each other. Another friend of mine, whom I met in the Maryland conference, was Ted Becker, an acquaintance from Berkeley, who worked on NMR spectroscopy at the National Institutes of Health. I have worked with Ted in many international committees and collaborated on studies of the hydrogen bond. Ted has also visited me a number of times in India.

I taught a course on infrared spectroscopy at Purdue, in the summer school. I could record the infrared spectra of several compounds of interest and also carry out some research with colleagues in the US. I collected lots of chemicals and labware, including heating elements and thermocouples for my research. I had a suitcase full of materials on my trip back to Bangalore. Fortunately, the customs authorities did not object to the little bottles filled with crystals and powders of different colours. I had bought some of the materials with my own savings.

Throughout my stay in the US, I worried about my scientific future. The real concern was how I would ever be able to carry out research in India at the level that I wanted to work at. I had failed to persuade the director of the Indian Institute of Science, Professor S. Bhagavantam, to spend around Rs 10 lakh to purchase some essential instruments I required. In the absence of any hope of procuring good equipment, I was at a loss. There were suggestions about taking up a faculty position in the US. But both my wife and I were not interested in shifting to the US. At that time, I heard that a new Indian Institute of Technology was being started in Kanpur with the support of a consortium of US universities, which included MIT, Berkeley, Ohio State, Michigan, Princeton and Purdue. Late in 1962, I was invited by IIT Kanpur for an interview and a few weeks later, I received an offer for associate professorship in chemistry. I arrived in Kanpur by train on a hot April day in 1963. Arriving at the Kanpur railway station from Bangalore can be a shock for anyone, but I decided to completely ignore all the difficulties and inconveniences that I was likely to face.

In early 1963, IIT Kanpur was functioning from a building in the city and had no campus. The campus itself was far away from the city and had no academic buildings or hostels. A few small houses (type III) had been built in the campus and I was given one of them to live in. I lived there and travelled to the institute in the city every day. It was difficult to sleep at night in the hot weather. There was no place to eat food in the campus. At lunch time, I had to eat samosas sold by vendors on the roadside. The temperature only kept increasing, reaching 40°C and above in May. I spent several weeks in this manner, often driving to the city to purchase things for the house and for the teaching laboratory. My main concern was about the new education programme for undergraduates, which was to start in July 1963. The workshop building was just getting ready and we were expected to set up undergraduate teaching laboratories there. The first student hostel was expected to be ready by June 1963. I was to teach the first course in chemistry (Chem 101) to students starting in July (as the instructor in charge) in a building that was just coming up. All the faculty members, independent of rank, were tutors for the course. We got the laboratories in the workshop building fixed by local carpenters and contractors. We had to ensure that the kind of experiments the students performed were interesting, and had to procure glassware and other laboratory requirements from various sources.

Two months after I arrived in Kanpur, my wife and daughter joined me. We were the first family to live in that isolated campus. Even to buy salt, we had to drive a few kilometres, but we did not mind these difficulties. We were looking forward to a great, new institution. Several young faculty members started joining the institute during 1963–65, most of them coming from the US or Europe. There were a few American professors from the consortium of US universities. They participated in discussions with regard to the curricula and other academic matters. In chemistry, we had Derek Davenport (from Purdue) and Truman Kohman (from Carnegie Mellon), both of whom were fine people and supported me and other colleagues fully.

Late in 1963, my book on infrared spectroscopy, published by Academic Press, New York, came out. I showed it to the institute's director, Dr P.K. Kelkar, who was delighted to see a book from Academic Press, where IIT Kanpur's name featured on the front page. While we faced a lot of physical discomfort in the early days at the IIT Kanpur campus, we all lived in great hope. The success of the institute was mainly because of the extraordinary aspiration of the faculty to build an

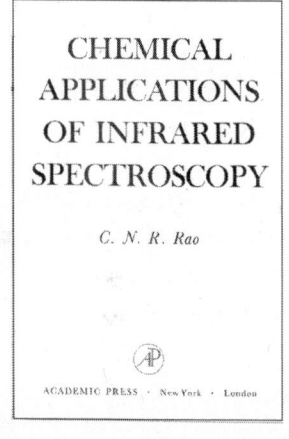

CHEMICAL
APPLICATIONS
OF INFRARED
SPECTROSCOPY

C. N. R. Rao

ACADEMIC PRESS · New York · London

institute of excellence. The campus was exciting because of the young faculty and enthusiastic students. Teaching was a pleasure. I was involved in the undergraduate programme in a big way and was convener of the programme. We had to decide on the kinds of courses to be offered and even on the method of evaluating the students. I can never forget my first two or three years at IIT Kanpur.

Along with the undergraduate programme, we in the chemistry department started PhD courses by July 1964 and an MSc programme a little later.

In dealing with the chemistry department, the director treated me as if I was the head of the department though he had not announced it. In July 1964, I was formally asked to be the head of the chemistry department and was promoted to full professor. I was just thirty years old and only IIT Kanpur could have made me head of a department. This was possible because Dr Kelkar was an outstanding director who trusted people like me. He had tremendous faith in the people who worked with him and gave us all the freedom and encouragement we needed. It was a pleasure to work with him. We also had a friendly deputy director, Dr

The P.K. Kelkar Library at IIT/Kanpur (Inset: Dr Kelkar)

M.S. Muthanna. I spent all available time in building the chemistry department and started recruiting outstanding young people from all over the world as faculty members. I had full cooperation from my senior colleagues, M.V. George and P.T. Narasimhan. We tried to teach modern chemistry of the kind that was being taught in the best institutions in the world. Even today, IIT Kanpur has the reputation as the institute which initiated the teaching of modern chemistry for the first time in India.

By the end of 1964, I had started doing some research at IIT Kanpur. That year also marks the birth of our son, Sanjay. The American aid programme provided us good infrared and UV-visible spectrometers, as well as X-ray diffraction equipment. The visiting American faculty helped in faculty recruitment and curriculum matters, as well as in getting visiting professors from the US. We could procure many items of equipment and chemicals from the US through the American programme. Unfortunately, the spectrometers were meant for common use in the department and I did not have access to a dedicated instrument for my own use. In spite of this, we could do reasonable work in optical spectroscopy, though not of the kind that I had planned earlier. Most of the work in spectroscopy that I did was, therefore, related to the study of hydrogen bonding and molecular interactions. I also carried out normal vibration analysis of several molecules. Unfortunately, I could not use NMR and EPR spectrometers, as they were under the control of individuals who turned them into their private possessions. In solid state chemistry, my work was on phase transitions in solids, defects in ionic solids and electronic properties of oxides. I got especially interested in metal oxides because they form the largest class of materials with a variety of structures and properties.

What I cannot forget about the early years at Kanpur was our determination to accomplish something worthwhile. Only the laboratories with sophisticated instruments were air-conditioned. Other laboratories, including my own, and lecture halls had no air-conditioning. We lectured 200 or more students in the heat of July–September, without paying heed to the weather. I remember writing manuscripts during between May–June, sitting under a fan on in my office. The fans circulated hot air, but our minds were cool. By the end of 1967, three of my students obtained their PhD degrees from IIT Kanpur. All my students at IISc had also completed their PhD degrees by then, the first three having received their degrees in 1962—three years after I joined IISc.

The financial situation of the chemistry department at IIT Kanpur was quite good, thanks, at least in part, to the supply of equipment and chemicals by the American programme. I also obtained two research grants from the Council of Scientific and Industrial Research (CSIR), providing two research fellows and a contingency grant of Rs 3000 per year. I then obtained fairly large grants from the PL 480 funds through the US National Bureau of Standards and the National Institutes of Health.

In July 1964, after my book on infrared spectroscopy came out, I received a wonderful letter from Professor C.V. Raman, in which he wrote that he was impressed by the book. He felt that the author of such a book should be a member of the Indian Academy of Sciences. He offered to nominate me and make sure I get elected. He kept his promise. Professor G.N. Ramachandran seconded the nomination. I can never forget this tremendous gesture from Professor Raman. Election to the fellowship of the Indian Academy of Sciences was the first professional recognition of my life.

SIR C. V. RAMAN
Director

RAMAN RESEARCH INSTITUTE
BANGALORE-6

Ref: No. 308   3rd July, 1964

Dear Dr. Rao,

The volume by you entitled "CHEMICAL APPLICATIONS OF INFRARED SPECTROSCOPY" published by the Academic Press has recently come into my hands. It is a praise-worthy effort that you undertook to write this book.

I feel that it would be an appropriate recognition of your fruitful scientific activity that you are elected as a Fellow of the Indian Academy of Sciences. This, of course, assumes that you would desire to be so elected and that you are prepared to fulfil the obligations of the Fellowship which include an entrance fee of Rs.50/- and an annual subscription fee of Rs.36/-. If you are so agreeable, necessary steps will be taken at this end. I enclose herewith a copy of the latest year book of the Academy which will give you any further information you may require.

Your full name, academic degrees and your official designation will be needed for proposing your name for election.

Encl: One book

Yours sincerely,

G. V. Raman

Dr. C.N. N. Rao,
Department of Chemistry,
Indian Institute of Technology,
KANPUR (INDIA)

Early in 1967, the Faraday Society, England, announced that I would be awarded the Marlow Medal. The recognition was meant for the most outstanding young physical chemist below the age of thirty-five. The award was for my contributions to solid state chemistry and spectroscopy. In September 1967, I went to Toronto with my wife to receive the Marlow Medal in a special meeting of the Faraday Society. Since I was going to north America, I thought I could also spend some time in the US to seriously think about the research areas that I should pursue in the years to come. IIT Kanpur allowed me to take one year off for this purpose and Purdue University offered me a visiting professorship. We had to leave our two children in Bangalore with Indu's parents to enable her to study for an M.S. degree in education at Purdue. Our hope was that this would help her broaden her interests. By then, she had obtained an M.A degree in sociology from Kanpur.

Family with Rao's parents

Family with Indu's parents

I collaborated with George Honig and others at Purdue, and my main accomplishments were in the area of metal-insulator transitions in metal oxides—a new area that was just emerging. Several oxides had been found to exhibit properties of metals like copper and I got interested in exploring transition metal oxides with such unusual properties. There were still very few practitioners of solid state chemistry at that

With Prof. George Honig

time—Paul Hagenmuller in France and John Goodenough in the US being prominent ones. I met Goodenough for the first time in a meeting in Boston in early 1968. We became friends immediately. Of the other faculty members at Purdue with whom I collaborated, I must mention Henry Feuer, an organic chemist, who remained a great friend until he passed away a few years ago. I taught physical chemistry and part of the general chemistry course during my stay at Purdue. I was thrilled when the students evaluated me as an outstanding teacher.

After returning from Purdue, I was asked to become the first Dean of research and development at IIT Kanpur. I took up this task and tried to set up proper structures for supporting research and development in the institute. E.C. Subbarao was the Dean of faculty. Subbarao and I started a postgraduate materials science programme. My stay at Purdue enabled me to decide that solid state and materials chemistry would be the main focus of my research in the years to come, rather than spectroscopy and molecular structure. I initiated research into many new areas including metal-insulator transitions and properties and phenomena exhibited by perovskite oxides. I kept busy doing research and teaching during this period, and also contributing to various aspects of the institute.

In 1969, the Indian Academy of Sciences celebrated Prof. C.V. Raman's eightieth birthday in Ahmedabad. Vikram Sarabhai was the local organizer and it was attended by the who's who of Indian science. I was one of the speakers. After my talk, Prof. Raman praised me publicly.

One day in 1970, I received an unexpected telegram from CSIR telling me that I had been awarded the Bhatnagar Prize for chemistry for the year 1968. The announcement of the award was two years late. I was not even aware that I had been nominated for the Bhatnagar Prize. This was a tremendous boost to my morale. Bhatnagar Prize awardees were expected to give a lecture on their work. I did so at Delhi University, with Professor T.R. Seshadri in the chair. I talked of new materials, especially metal oxides, and their amazing properties. Unfortunately, the chairman made fun of the kind of chemistry I talked about. I had to keep quiet. I thought to myself then that I would never ridicule a young scientist.

**Rao receiving the Bhatnagar Prize from
Mr C. Subramaniam (1968)**

From 1963 to 1970, I shared a close working relation with Dr Kelkar. He was a director in pursuit of excellence, and took pride in the institute and its faculty. He would often call me to his office around 5 p.m. and talk to me for extended periods of time, much to the consternation of my wife. During such meetings, he used to tell me, 'You should become a FRS (fellow of the Royal Society). I have the feeling that you will.'

In 1970, Dr Kelkar retired as the director of IIT Kanpur. He was the architect of the golden period of the institute. I was asked to take up the directorship by the minister of education in Delhi, but I had no interest in doing so. I had many things to do in science. I told the minister that I was just thirty-seven years old and did not want to be director of an IIT. He was surprised. Dr Muthana became the director. For various reasons, many problems arose in the campus soon after.

I attended the 1970 meeting of the Indian Academy of

With Dr Muthana at the convocation, IIT Kanpur (1967)

Sciences, at the specific invitation of Prof. Raman. The meeting was in the premises of the academy in Bangalore. There were very few participants, but Prof. Raman was sitting in the front row. I gave a talk on the structure of water. He was eloquent in his praise after the lecture. A few weeks later, he passed away. We had lost a giant in science.

One of the most important efforts I made during the 1969–74 period was to organize winter schools in solid state chemistry, with experts from abroad such as John Goodenough, J.S. Anderson and George Honig. The schools were a great success. I also edited a book titled *Modern Aspects of Solid State Chemistry* based on the lectures at the first school. The book, published by Plenum Press (1970), was the first (nearly) complete book on solid state chemistry. There was another book on the subject edited by me, and published by Marcel Dekker, based on the 1974 winter school.

In 1971, Prime Minister Indira Gandhi decided to set up a National Committee for Science and Technology (NCST) with C. Subramaniam (CS) as the chairman to chalk out an effective science and technology plan for India. CS was the minister for science and planning in the Government of India. There were nine members in the council and I was asked to be one of them. Some of the other members included M.S. Swaminathan, Raja Ramanna and

Smt. Indira Gandhi

Dr Arcot Ramachandran

V. Ramalingasami. Working with CS was a great experience for me. One of the first recommendations of the NCST was to establish a new Department for Science and Technology (DST). Soon, my dear friend Arcot Ramachandran (then the director of IIT Madras), became secretary of the department and contributed much to the areas of science and technology in India. In the NCST, I was to take care of basic science, science education and advanced materials. Some of the recommendations made by us became realities. The setting up of the science and engineering research council and regional sophisticated instrumentation centres are two such examples. A recommendation to set up national science universities similar to IITs did not fructify.

NCST came up with recommendations on several aspects of science and technology. The plans and programmes proposed by the NCST were outstanding and the document produced by the council remains the best and the most complete plan for science and technology in India. In preparing the plan, a few thousand scientists were involved. In addition to the rewarding experience of working in the NCST, I also gained first-hand knowledge of industries as the chairman of a small public sector chemical industry (Hindustan Insecticides) between 1969–73. On 26 January 1974, I was pleasantly surprised to learn that I was being awarded the national honour, the Padma Shri. I had no idea that I had even been nominated. Clearly, it must have been a gift from CS. During the NCST days, I enjoyed working with Arcot Ramachandran. I was also involved with the working of the new Science and Engineering Research Council, chaired by Arcot Ramachandran and administered by R.D. Deshpande. During our frequent meetings, Arcot used to tell me, 'CNR, you should get those three letters.' The three letters referred to FRS. I would wonder whether it would ever happen.

During the 1970s, I was deeply involved in teaching chemistry to undergraduates. I took immense trouble to make the first course in chemistry to undergraduates (Chem 101) an exciting experience and I enjoyed doing so. Even today, when I meet any of the few thousand undergraduates I taught then, the first thing they remember is my Chem 101 course. I have had heart-warming experiences in this regard. When I was in Ras al-Khaimah (RAK) in the United Arab Emirates for a symposium on materials, I was in conversation with the ruler of the Emirates, Sheikh Saud. A tall gentleman came in and the Sheikh introduced him as his financial and business adviser. The next thing I knew, the adviser had prostrated himself on the floor before me, touching my feet. He got up and said, 'I will never forget your Chem 101.' When I was in Australia to be admitted to the academy as a foreign member, a reception was organized by the Indian community. Just as I entered the reception hall, two middle-aged gentlemen rushed to me and asked, 'Do you remember us from Chem 101?' I have had several such encounters all over the world—in airports and elsewhere. I was in the national Chemical Research

Society meeting recently, where a distinguished professor of Indian origin from the US was to give a plenary lecture. He started his lecture stating that he owed much to my Chem 101 course at IIT Kanpur and the advice that I gave him subsequently about his career.

I wrote a book *University General Chemistry* along with a few colleagues for freshers in college. The book, published in 1973 by Macmillan, presents all the important aspects of chemistry in simple language. It has many examples and solutions to problems. The book is still selling and is widely used by young people looking for admission in leading institutions. I also authored the book *Experiments in General Chemistry*. I must mention that in the practical class for freshers, we came up with experiments that would interest them. One of them was to determine the acidity of Coca-Cola, where they could drink all the Coke that was unused. Students were also asked to prepare aspirin tablets. They enjoyed tablet making more than the chemical preparation of the compound.

In 1973, I was asked by the committee on teaching of chemistry of the International Union of Pure and Applied Chemistry (IUPAC) to become a member. I started working with them in earnest. I was also involved in national programmes in science education, led by Professor D.S. Kothari and lectured to chemistry teachers all over India. With the help of some colleagues, I used to produce a science newsletter, which was mailed to 5000 school and college teachers. Students of the Opportunity School, headed by my wife, took care of mailing the newsletters. The Opportunity School

Rao at the Indian Institute of Technology, Kanpur

was organized in the campus to provide good education to first-generation learners, generally the children of daily wage workers. In 1973, I was awarded the Jawaharlal Nehru Fellowship for a period of two years and I gave up the Deanship of the institute.

My heavy involvement in research and teaching at IIT Kanpur, and my work with the NCST kept me really busy. My day would begin as early as 7.30 or 8 a.m. and go on till 7 p.m. every day of the week. Life in the campus was enjoyable, with lots of dinner parties. We also entertained friends and colleagues quite frequently at home. Scientists visiting IIT Kanpur from other places inevitably ended up coming home. IIT Kanpur had emerged as the best institute of higher education in science and engineering by the 1970s, and the reputation attracted

With Prof. M.V. George at IIT Kanpur

many scientists to the campus, in spite of the fact Kanpur was not the easiest place to reach. IIT Kanpur was the envy of other institutions. It is not an exaggeration to state that some of the finest chemists one sees in the country today are the alumni of IIT Kanpur. Some of the faculty members from here, who have become well known are D. Balasubramanian, Animesh Chakravorty, G. Mehta and S. Ranganathan, besides my senior colleagues George and Narasimhan.

I was elected fellow of the Indian National Science Academy in 1974. More importantly, I received a letter from Oxford University in June 1974, inviting me to be a Commonwealth Professor in chemistry, offering me the salary of a professor for one year, and travel support for me and my family to go to Oxford. I was supposed to start in October 1974 at the Inorganic Chemistry Laboratory (ICL) at Oxford. Since I was not expecting such an offer, it took some time to get used to the situation. The chairman of the board at IIT Kanpur, Dr Hussain Zaheer, and the director, Dr Muthana, encouraged me to go to Oxford. They considered the invitation from Oxford a great honour.

By the end of September 1974, I went to Oxford with my wife and two children, Suchitra and Sanjay. I went to ICL the day after I reached Oxford and started exploring ways to initiate research. We lived in a nice house in a village outside Oxford, called Old Marston, and my children went to a school nearby. Indu joined the department of education in Oxford, to take courses in psychology and other subjects.

Oxford University

Our stay at Oxford was fantastic. The inorganic chemistry laboratory, headed by Professor J.S. Anderson, was probably one of the

Oxford University

best centres for solid state chemistry in the world. There were people working on almost all aspects of solid state chemistry, including electron microscopy, various spectroscopies, neutron diffraction, electronic structure, synthesis and so on. I decided to learn as much as possible by working with as many people as I could. Indu used to make fun of me that I was a foreign student in Oxford and not a visiting professor. She was right. I almost became nervous at one stage, working day and night making compounds and studying phenomena and properties with several Oxford colleges.

I remember an incident from the Christmas holidays in 1974. The laboratory turned off the heat at that time and one had to wear a top coat inside the room if one wanted to be in the building. I got bored of sitting at home and decided to go to my office to write some manuscripts. I suddenly heard some noise from the attic. I walked up to the attic to see what was happening. There he

Prof. J.S. Anderson

was, Professor J.S. Anderson working at a table, wearing a top coat. He was analysing gallium in one of the compounds since his student had some difficulty. It was heart-warming to see old Professor Anderson working in the lab the day before Christmas.

When my year at Oxford came to an end, I had nearly seventeen papers with Oxford faculty members, including Professor Anderson. I did high resolution electron microscopy on Aurivillius phases and other materials, and published a few papers on the subject with Professor Anderson and others. I worked on X-ray photoemission spectroscopy of solids (with Tony Orchard) and UV photoemission spectroscopy of molecules (with Jenny Green). I had papers on the Raman spectra of oxides as well.

During my time at Oxford, I got to meet Professor Nevill Mott, who had come from Cambridge to give a seminar. Professor Mott is one of the main founders of modern solid state science. The story of his Nobel Prize is interesting. For reasons that many of us do not understand, he was not awarded the Nobel Prize for his monumental work on various aspects of condensed matter while he was Cavendish Professor of Cambridge. He then took voluntary retirement at the age of sixty-five from the Cavendish professorship, and started work on an entirely new problem—electrons in

**Sir Nevill Francis Mott
—Rao's Guru**

random lattices—for which he received the Nobel Prize for that work. I started collaborating with Professor Mott on strongly correlated oxides and visited him in Cambridge a few times. He was simple and quite a gentleman. I published some work with him as well. With Tony Cheetham, I worked on a neutron diffraction study of NbO_2.

By the end of September 1974, I had to come back to IIT Kanpur. My wife and I had enjoyed our stay at Oxford, both socially and academically. My association with St Catherine's College as a professorial fellow gave me the experience of college life and the pleasures of the high table. Many of my Oxford colleagues, including Professor Anderson, used to visit us at home. Our children also liked their stay in Oxford. Indu received a certificate in education from the university's department of education. Towards the end of my stay, I started planning a book on phase transitions in solids.

Returning to India from Oxford raised some difficult questions in my mind. These were issues related to the sophisticated instruments and other facilities that I would not have when I came back to Kanpur. I decided that I must do my best to establish a laboratory somewhere in India, which had all the necessary facilities to carry out frontier research in solid state and materials chemistry. Once this idea entered my head, it became clear I would have to leave IIT Kanpur. It was not possible to do the kind of research that I had in mind in Kanpur at that time or set up a laboratory required for the purpose. Part of the reason was the disharmony that prevailed in the campus among the faculty and the problems with employees after Dr Kelkar's departure in 1971. One solution would be to leave India and go to US or somewhere else. I debated the pros and cons of the various possibilities before me.

I had an unusual experience during this period. Prime Minister Indira Gandhi wanted to meet me in her house. I went there early in the morning. I was treated to hot coffee and then she appeared, fresh and elegantly dressed. She said, 'I am looking for young people to head scientific agencies. I would like you to be a secretary to the government.' For a minute, I was in shock. I soon recovered, and told her that I did not want to give up research and become a full-time government employee. She then remarked, 'Are you sure? You are the first person to have refused such an offer.' She shook my hand and I left the room.

I was going through a miserable time, not knowing what to do. I found it difficult to leave IIT Kanpur, which had done so much for me and which I loved. I established myself as a scientist only after going to Kanpur. My family and I had enjoyed our stay on campus. Then, there was also the problem of finding a really attractive alternative place in India, and to obtain major funding (of at least Rs 1–2 crore). I did not have any idea how I would do this. During this period, I received a medal from the American Chemical Society. When I went to receive the medal in New York, in September 1976, the Deans of some of universities discussed my relocating to the US with me. The function itself was wonderful, with Glen Seaborg presenting the medal and Linus Pauling giving the centenary lecture. The occasion was used to wipe out the bad history of the society with relation to Pauling.

While I was still worrying about my professional future, I met Satish Dhawan at a meeting of the Indian Academy of Sciences. He asked me, 'Why do you have to leave India and go to the

Prof. Glen Seaborg presenting the American Chemical Society Medal (1976)

US? What can I do to get you come back to Bangalore?' I told him of my desire to set up an outstanding department for solid state and materials chemistry. Satish suggested many ways he could accommodate me in the Indian Institute of Science. He wanted me to head the chemistry division, but I was averse to that suggestion. After some discussion, it was agreed to set up a new department devoted to solid state and structural chemistry at IISc. Unfortunately, he could not provide any financial support for the proposed new department, and I had to find the funds to procure equipment and set up the various facilities. I decided to take up the task.

My decision to move back to Bangalore from Kanpur cheered up mine and Indu's parents. I shifted to IISc in November 1976. Fortunately, the Science and Engineering Research Council of the Department of Science and Technology gave me two major grants which amounted to Rs 60 to 70 lakh. In the meantime, IISc received a grant to set up a materials research centre. Professor Dhawan asked me to head the materials research centre as well. I started to build a new department named Solid State and Structural Chemistry Unit and also the new Materials Research

With Prof. Satish Dhawan at IISc

Centre within a few weeks of joining the institute. The idea of calling the new department a 'unit' was Satish's. He had set up a molecular biology unit with Professor G.N. Ramachandran a few years earlier. I needed some immediate laboratory space to start research work, but the chemistry division was blunt when they told me they could not provide even a single room. Fortunately, the chemical engineering and metallurgy departments gave me a few rooms

Prof. G.N. Ramachandran

to set up my laboratories. Some of my students from IIT Kanpur and new ones recruited from IISc helped me initiate research, making use of the equipment and chemicals that I had brought with me from Kanpur.

I placed orders for a good infrared spectrometer, a UV-visible spectrometer, an X-ray photoelectron spectrometer, an EPR spectrometer, and so on. I also placed orders for a laser Raman spectrometer. The institute would have a Raman instrument for the first time. I could get both scanning and transmission electron microscopes and an X-ray diffractometer for the Materials Research Centre. For the first time, I had reasonable facilities for research by mid-1977, eighteen years after I started my independent career. I proposed to carry out research of the kind that I had always wanted to do. I made sure that there was a dedicated diesel generator to ensure constant supply of electricity. The city electric supply has been unreliable for as long as I remember and the situation is still the same even today.

Shifting to Bangalore caused a great deal of financial problems for me. We had a fine house in the campus at Kanpur.

IISc did not give me a house to live in and I had to spend a major part of my salary on the rent for a house in Malleswaram, not far from the institute. Then there were difficulties in getting the children settled, and finding the right school and college. Building the new Solid State and Structural Chemistry Unit was, however, a matter of great pleasure. I was able to recruit a few young faculty members and, with the new equipment, I could start research in some new areas. We also started teaching several new postgraduate courses, and organized various types of seminars and mini symposia, besides winter schools in the chemistry of materials. The new unit soon gained a good reputation and became the envy of others in the institute. The unit became prolific in research in a short time and was publishing the best of papers in the best of journals. The unit established itself as an important centre of research in solid state chemistry internationally. The Materials Research Centre provided the much-needed facilities for research to all those working in areas related to solid state and materials science. In 1978, my book *Phase Transitions in Solids*, co-authored by K. J. Rao, was published by McGraw-Hill, New York.

With Prof. Ramaseshan

After Professor Raman's death, I was closely involved with the Indian Academy of Sciences and worked closely with S. Ramaseshan. I became the first scientific secretary of the academy and did much to organize various activities, especially the annual meetings. Several new programmes were initiated, including topical symposia, creation of the young associateship scheme and so on.

I was secretary for two terms, and later the editor of publications and Vice President. We reorganized the journals and started a new journal for materials science. The chemistry journal, was made a full-fledged journal independent of the proceedings. I was the first editor of these two new journals.

My main preoccupation during the period between 1977–82 was to carry out the best possible research in solid state and structural chemistry from Bangalore. This was possible to a great extent with the facilities available, although there were some limitations. I worked mainly on the chemistry of materials, particularly metal oxides, but had a few students in spectroscopy and structure. I had lots of wonderful students working with me, and we built photoemission and energy loss spectrometers to study the electronic structures of solids as well as of free molecules. We published several papers on photoemission spectroscopy at that time. I made some effort in helping the Department of Science and Technology with the research council, and was a member of the University Grants Commission. I continued my involvement with activities related to science education and had, by that time, become chairman of the Committee on the Teaching of Chemistry of International Union of Pure and Applied

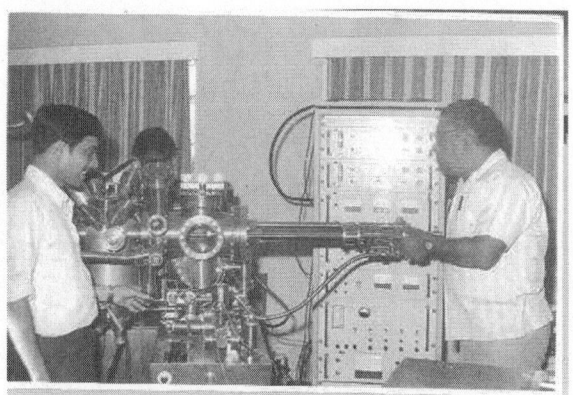

Rao at the Solid State and Structural Chemistry Unit, Indian Institute of Science

Chemistry (IUPAC). I could involve outstanding chemists and chemical educators in the work of the committee, especially in the annual international conferences that I initiated. I was also a member of the IUPAC spectroscopy commission then. In 1977, I was elected member of the bureau of IUPAC. The head office of IUPAC was in Oxford and the bureau met once a year. My main preoccupation, however, remained research and teaching.

Suchitra's wedding day (1981)

In October 1981, my daughter, Suchitra, married Dr K.N. Ganesh, a chemist with PhDs from Delhi and Cambridge. It was a happy occasion. A little later, in 1982, I learnt that I had been awarded the Solid State Chemistry Medal by the Royal Society of Chemistry. Around the same time, Purdue University wrote to me that it would like to confer the honoris causa doctor of science degree on me in the June 1982 commencement. I felt elated that my alma mater in the US was the first one to confer this honour. My wife and I decided to go to Purdue University in 1982 for the ceremony and spend a few weeks there. I was also scheduled to give a talk in a symposium of the American Chemical Society at Las Vegas in the last week of May 1982.

On our way to the US, my wife and I stopped in Paris for three to four days to visit some friends. When I was in Paris, I learnt that I had been elected a fellow of the Royal Society. I knew that I had been nominated by leading chemists in the UK for the fellowship, but did not expect that I would be elected so soon. This was a great moment, and made me feel humble and grateful. I had not expected to be a FRS and wondered

PURDUE UNIVERSITY

UPON THE NOMINATION OF THE PRESIDENT AND THE FACULTY
AND BY THE AUTHORITY OF THE TRUSTEES

HAS ADMITTED

C. N. R. RAO

TO THE DEGREE OF

DOCTOR OF SCIENCE

FOR HIS PIONEERING CONTRIBUTIONS TO MANY
AREAS OF SOLID STATE AND STRUCTURAL CHEMISTRY

WEST LAFAYETTE, INDIANA
MAY 15, 1982

PRESIDENT OF THE TRUSTEES

Arthur G. Hansen
PRESIDENT OF THE UNIVERSITY

C. N. R. RAO   DOCTOR OF SCIENCE

In a letter to the New York Times in 1945, James Bryant Conant, the president of Harvard University, declared, "There is only one proved method of assisting the advancement of pure science — that of picking men of genius, backing them heavily, and leaving them to direct themselves."

Certainly, Conant's observation is all the more true because of the example of Dr. C. N. R. Rao, who is one of chemistry's most prodigious researchers — a man of genius — in the complexities of chemical spectroscopy, molecular structure, and solid-state chemistry.

He is world-renowned and often honored in his field for his impressive contributions to the world's scientific literature which now include 10 books and more than 300 research papers. Most recent of the many prestigious honors that have come to him was his being named a Fellow of the Royal Society in London.

Professor Rao's honors and awards are impressive and include, besides membership in the prestigious Royal Society, the Marlow Medal of the Faraday Society; S. S. Bhatnagar Prize in Chemistry; Sri C. V. Raman Award in Physical Sciences; Centennial Foreign Fellowship of the American Chemical Society; Federation of the Indian Chamber of Commerce and Industry Prize for Research in physical sciences; and the S. N. Boss Medal of the Indian National Science Academy. The president of India bestowed upon him the national honor, Padmashri, in 1974. He was elected a foreign member of the Academy of Sciences of Yugoslavia in 1981.

He is also distinguished by the fact that his wife, Indumati, is an alumna of Purdue, having received the M.S. degree in 1968.

Purdue is proud to count Professor Rao as one of its own.

Honorary fellowship of the Royal Society of Chemistry, London

whether I had proved myself to deserve the honour. My parents were delighted.

We returned to Bangalore after I received the DSc (honoris causa) from Purdue and attended a Gordon Research Conference in New Hampshire. My election to the Royal Society strengthened my resolve never to give up research.

Between 1970 and 1982, I attended many spectroscopy conferences. My membership (later chairmanship) of the IUPAC Spectroscopy Commission was partly responsible for this. My friends, Orville-Thomas and Ratajczak, were involved in organizing many of the meetings. Orville-Thomas had a large following and he was the leader of a group not too close to some of the prominent spectroscopists like Professor H.W. Thompson (Oxford). People were somewhat afraid of Prof. Thompson, who was foreign secretary of the Royal Society. I got to know him fairly well. He was also President of IUPAC and editor of a research journal devoted to spectroscopy. Orville also

Prof. Henryk Ratajczak with Mrs & Prof. Orville-Thomas

started many journals dealing with structure and spectroscopy. I was on the editorial boards of the journals edited by both these men. Personality clashes amongst spectroscopists arose because of competition or historical reasons. For example, there was great rivalry in the early days between the Oxford and Cambridge groups in infrared spectroscopy. Both built IR spectrometers independently around the same time. A leading spectroscopist like Mansel Davies did not get due recognition because he was far away in Aberystwyth in Wales and also because he was far too outspoken. Orville-Thomas was a student of Mansel Davies. I still miss Orville.

In August 1983, I received an invitation from Cambridge University, to be the first Jawaharlal Nehru Professor starting October 1983. I had also been elected professorial fellow of King's College. I decided to go to the chemistry department at Cambridge after taking a sabbatical from the institute. Prime Minister Indira Gandhi called to congratulate me on becoming the first person to occupy the professorship named after her father. When she called me, I was in a hospital with my wife, who was undergoing some tests. Her office had traced me to the hospital. Such personal interest from a Prime Minister does a lot of good to one's morale.

In 1983, University of Bordeaux and Sri Venkateshwara University conferred honorary DSc degrees on me.

In closing

The period between 1959–83 was a trying period in my life, when I was trying to prove myself the whole time. It can also be considered as a period of apprenticeship. Whatever one may consider this period to be, it was certainly one when I spent many a sleepless night worrying about my work and my professional future. Doing great science in India had become the main goal of my life. By 1983, it appeared as though I may have been somewhat successful in this effort.

'A man's reach must exceed his grasp;
Else what's a heaven for'— Robert Browning

5. THOSE BUSY DAYS: WORKING AGAINST ODDS

I went to Cambridge towards the end of September 1983 to take up the Nehru Professorship. I found a nice house on Chaucer Road, close to the botanical gardens and not far from the chemistry department. I was able to start some

King's College, Cambridge

research with my friend Professor John Thomas and other colleagues within a short time. I thought that I should start writing a book *New Directions in Solid State Chemistry* during my stay at Cambridge. The book was intended to present the emerging trends in this important subject. Interestingly, solid state chemistry, which was not a mainstream topic in chemical science in the 1950s, had become an area of vital interest by the 1980s. In its new avatar, it is called 'materials chemistry'. I started working on the book with

Sir John M. Thomas

J. Gopalakrishnan as the co-author. It was to be published by the Cambridge University Press.

A significant aspect of my stay in Cambridge was my close association with Professor Nevill Mott, who was about to turn eighty then. Along with Peter Edwards from the chemistry department, I edited a book as a Festschrift to celebrate Professor Mott's eightieth birthday. The book dealt with various aspects of Professor Mott's favourite subject, metal-insulator transitions. This was published by Taylor and Francis (1985).

My stay in Cambridge between 1983–84 was enjoyable, although I had started feeling somewhat unwell and had developed high blood pressure. My association with King's College made the stay socially wonderful. Martin Rees, the famous cosmologist and astronomer, was one of my close friends at the college. I also got to know Sydney Brenner, Fred Sanger and many others who frequented the college. Suchitra and Ganesh came to Cambridge for a few months and stayed with me, while Ganesh was working at the MRC laboratories. I published several papers with John Thomas and other members of the faculty, and gave lectures in several universities in Britain. In 1984, the Royal Society of Chemistry conferred its honorary fellowship on me.

I was to return to the Indian Institute of Science around September 1984. In June 1984, I received a phone call from Professor Satish Dhawan to inform me that I was being offered the directorship of IISc. He wanted me to take up the offer. He felt it was important that I do so. J.R.D. Tata, President of the Court of IISc, was also keen that I became the director. I did not know what to do since the directorship of IISc was not part of my plans. I returned to Bangalore in a confused state of mind. People were waiting for me to take charge as the director of IISc.

IISc needed quite a bit of doing. It required major financial inputs for improving its infrastructure. There was an urgent

With Mr J.R.D. Tata, President of the IISc court (1980)

need to improve facilities in the laboratories and offices. More student hostels and housing for faculty and support staff were also needed. Good research equipment was probably the greatest requirement of all. The electricity and water supply in the campus were terrible. Its plumbing was almost seventy years old and most of the laboratories did not get water for experiments because of clogged pipes. I thought that I would first concentrate on the infrastructure of the institute and then do the other things gradually. I was then chair of the commission on spectroscopy of IUPAC and had just been elected as Vice President (the next President) of the IUPAC around that time. I was also Vice President of the Indian National Science Academy (INSA) in Delhi. I had much to do outside my research, but there was no way I was going to sacrifice my research. I had to find a new style of working and a way to manage time. Time management is clearly one of the most crucial things in one's life if one wants to accomplish anything worthwhile.

IISc was seventy-five years old when I took over as director and we had to celebrate its platinum jubilee. I could get Indira Gandhi to agree to come for the platinum jubilee function in the campus. Unfortunately, before she could come, she was

assassinated in October 1984. We then invited the Vice President
of India, R. Venkataraman, as the chief guest. In January 1985,
we celebrated the platinum jubilee at the campus. I managed to
get the history of the institute written by Dr B.V. Subbarayappa,
called *Pursuit of Excellence* (published by Tata McGraw-Hill).
The book recorded historical facts about the institute and
provided a status report as of 1985. It had a foreword written by
J.R.D. Tata.

Early in the
morning on 26
January 1985, I was
surprised to receive a
telegram informing
me that I had been
awarded the national
honour of Padma
Vibhushan by the
President of India. I
thought that it was
a typing mistake,
and that it must be
the Padma Bhushan.
But it was not so;
it was indeed the
Padma Vibhushan.
I later learnt that
Indira Gandhi was
particular that I

**Receiving the Padma Vibhushan from the
late President, Zail Singh (1985)**

receive the Padma Vibhushan.

By January 1985, my health seemed to be deteriorating and
I found it difficult to even walk sometimes. I consulted my dear
friend Dr K.G. Nair, a cardiologist in Bombay. Dr Nair is one
of the most outstanding doctors I have known. He felt that my
health required immediate attention. Since I was going to London

in February 1985 to give a talk at a Royal Society symposium, he felt I should consult doctors at the Harley Street Clinic. I decided to follow up on his advice. He had contacted the doctors in advance. Doctors at the Harley Street Clinic advised me that I should immediately undergo open heart surgery. I agreed to do so and took two weeks of rest in Cambridge, at Robinson College, after the surgery. I returned to Delhi a day before the date set for the Padma award ceremony at the Rashtrapati Bhavan. I received this honour from the President before going back to Bangalore. Nobody in Bangalore knew about the surgery. I informed my parents that I had to undergo a minor surgery only after I returned. I started work at the institute the following week.

I approached the Planning Commission for a special grant for improving the basic infrastructure at the Indian Institute of Science. Thankfully, they gave me Rs 3 crore so that we could set up a new electric sub-station and change the plumbing on campus. One of the minor contribution I could make, in the meantime, was to stop the grazing of cattle in the campus. The cattle belonged to employees who allowed them to roam freely around the campus. It was difficult, but I had to stick to the decision. There were minor demonstrations by the employees, but they eventually calmed down. They realized that the IISc campus was meant to have gardens for people to walk around in, not for cows to graze. With patience, I could also stop people from parking their bicycles in the corridors of the main building, effectively blocking the way. I started improving the gardens by planting several thousands of trees. We had to plant a large number of trees in front of the main building so as to cover the trees planted in the form of a Union Jack by the first director of the Institute, Morris Travers. At the same time, I instituted an annual lecture named after Travers. The more important contributions that I had to make related to academic matters.

As I was dealing with the directorship of IISc, I was very concerned about my research. How could I carry out good research if the directorship took up most of my time? This question continued to nag me. At that time, I also had to work for IUPAC and INSA. My working hours, therefore, had to be different. I decided to get up very early in the morning, go for a brisk walk and reach the laboratory at 8 a.m. I worked in the laboratory till about 10.30 a.m. and then went to the director's office. I would come back to my department around noon and work there till about 3 p.m. before returning to the director's office. I would come back to the laboratory again at 5 p.m. or earlier. I worked seven days a week then (and still do).

I could manage to do research in some areas I was interested in, but it was not easy. It was difficult to decide which of the demands on my time should receive high priority. I would arrange tasks in my mind in a particular order and allocate the time required for each demand in a suitable manner. This was not always satisfactory.

A wonderful event in my personal life at this time was the birth of our first grandchild, Kartik, on 14 November 1985.

As mentioned earlier, facilities in most departments were poor. For example, there was no photocopier in any department. One of the first things I did was to make sure that every department had its own photocopier and related facilities. Improving the phone system was another major task at hand. I urged each department to apply for funding to improve infrastructure, since such funding was available from the University Grant Commission and the Department of Science and Technology. I made sure that every department in the institute applied for such support. Within the next couple of years, most departments received funds for improving infrastructure from the Committee for Supporting Infrastructure in Science and Technology (COSIST). I initiated a major drive to recruit bright young faculty in every department. Some of the departments threatened me by saying it would not be possible for me to recruit faculty for their departments because of conflicting views. I did so in the case of biology by opening new sections and departments devoted to life sciences. Soon, the old departments wanted new faculty members as well. I could recruit nearly 100 young faculty members in about four years. In retrospect, many of those recruited to the institute then have distinguished themselves as leaders in their subjects. I also made sure that accomplished faculty members were nominated for appropriate recognition.

Rajiv Gandhi became the Prime Minister in 1985. Soon after, he inaugurated a conference of scientists and technologists in Delhi, organized to discuss national priorities. After the conference, Gandhi decided to establish a small Science Advisory Council to the Prime Minister. The idea was to have a council with a small number of members who would work in their personal capacities and not because of their positions in government. Rajiv Gandhi sent for me while I was attending the conference and asked me to take up the council's chairmanship. Before I knew it, my chairmanship of the council had been announced. Ashok Ganguly, Roddam Narasimha, R.A. Mashelkar. P. Rama Rao, M. Gadgil, Sam Pitroda, P.N. Tandon and J. Narlikar were some of the members. P.J. Lavakare was secretary of the council. My tenure as the chairman was enjoyable, though demanding. Gandhi would ask questions on various matters and would need quick answers. In addition, as a council, we wanted to make important recommendations on various aspects of science and technology. All this took time and effort. I was President of both IUPAC and INSA at that time, and held four important positions. I was also associated with the Committee on Data (CODATA) of the International Council of Scientific Unions (ICSU). Doing an honest job of any or all these functions, and carry out research at the same time, was a major dilemma I faced. I found it difficult to do the right thing at the right time. It became increasingly difficult to pay the required attention to ensure quality in my research. I tried to lessen my burden by giving up some responsibilities, one of them being CODATA.

My book *New Directions in Solid State Chemistry* was published in 1986 by Cambridge University Press. It was received well and soon got translated into Chinese and Russian. The book has since been reprinted and two different editions have also come out.

While I was working against time every day, there was a major development in science. In December 1986, a conference was organized at Hotel Ashok in Bangalore on valence fluctuation and I was to deliver one of the main lectures. Just before my lecture, Professor P.W. Anderson from Princeton University gave a lecture. When I finished my talk, Professor Anderson rushed to me and asked to speak with me urgently. He

asked me whether I knew about the discovery of high temperature superconductivity in an oxide. The oxide had been found to be superconducting at 35K, surpassing the 23K limit that existed till then. I did not know about this. There was no internet in those days and no simple way of learning about developments elsewhere in a short period of time. We depended on printed journals and the postal system. I asked Professor Anderson what exactly the nature of the oxide material was. He mentioned that the oxide had lanthanum, copper and so on. I asked him if it was by any chance La_2CuO_4? He said it was something like that. He then came with me to my office at IISc and spent two hours going through all the work that I had done on La_2CuO_4 since the 1970s. I had shown that it was antiferromagnetic. Unfortunately, I had not measured properties of La_2CuO_4 compositions at liquid helium temperatures due to paucity of facilities. The area of high temperature superconductivity soon became a rage in physical sciences.

By early 1987, everybody was talking about superconductivity. It was clear that I had to do something in this area. I started some work on materials similar to $La_{2-x}Sr_x(Ba)_xCuO_4$. Soon, there was a rumour that somebody had made an oxide which was superconducting at liquid nitrogen temperature, unlike the lanthanum compound which became superconducting at liquid helium temperatures (35K). The new liquid nitrogen superconductor was supposed to have rare earth, barium, copper and so on. The approximate formula had been given, but the exact composition or structure was not described. Apparently, there was a publication submitted to *Physical Review Letters* by Chu and Wu from the US. Everybody in the world wanted to get into this area and make the first well-characterized liquid nitrogen superconductor. I frantically started working on this in my laboratory. We worked day and night, and eventually made a compound which we thought was the right one. It showed a resistivity transition at 90K and had the composition $YBa_2Cu_3O_7$.

We could not measure the magnetic properties in my laboratory, but fortunately, Arup Raychaudhuri of the physics department could measure AC susceptibility and establish diamagnetism of this cuprate below 90K. I still remember that early morning when I drove to the physics department to see the result. I rushed a communication to *Nature* which was accepted. I then realized that we had made the first liquid nitrogen superconductor independently. Bell Laboratories in the US and Peking University in China had also prepared this compound at the same time. Those

Rao with some of his colleagues who worked on superconductivity

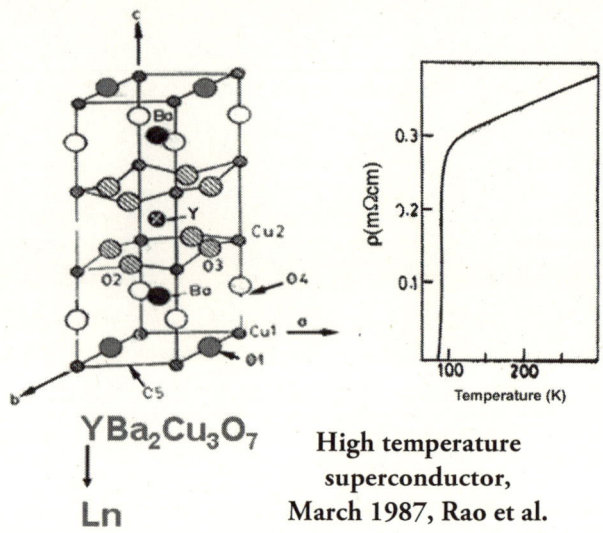

YBa$_2$Cu$_3$O$_7$

$\downarrow$

Ln

**High temperature
superconductor,
March 1987, Rao et al.**

were mad days. While it was enjoyable to be part of the gold rush,
it was also exhausting mentally and physically. The directorship
often came in the way of my research. Many of my ideas could
not be executed in time. It was common to find that somebody
somewhere had done exactly what I had planned. This was the
case with the bismuth copper oxide superconductors. We did,
however, carry out research on many aspects of high temperature
superconductivity and made some significant contributions. The
problem of being an administrator was that it became extremely
difficult, and at times impossible, to do competitive research.

I had a curious experience during the busy days spent
working on superconductivity. My friend, Ramaseshan, asked
me to publish one or two articles on the subject in the journals
of the Indian Academy of Sciences. By that time, I had papers
in *Nature, Physical Reviews* and so on. I sent two short articles to
be published in the academy journals. A few months later, some
'friends' commented that these papers had no date of receipt and
suspected un-academic behaviour. When I asked people at the
academy office, they expressed their regret for not printing the

dates of receipt and sent me a letter of apology for the oversight. It was later found that papers of a few others had also been published without dates of receipt. The problem ended soon afterwards.

By 1989, I had completed five years of directorship. It was also the year of the birth centenary of Pandit Jawaharlal Nehru, India's first Prime Minister. I wanted to give up the directorship of IISc. That was the only way to ensure I could continue doing research without a fall in its quality, even though I might publish some papers. I requested to be relieved of the directorship, but there was pressure from the faculty, the chairman of the council (Dr Raja Ramanna) and others that I should continue for at least another term. That meant being director of IISc for another five years, from 1989 to 1994. I was given no choice and I had to continue as the director of IISc.

In order to celebrate the centenary of Pandit Nehru, a national committee had been formed with R. Venkatraman as the chairman. The Science Advisory Council was asked to suggest ideas to this committee. We suggested the possibility of setting up new scientific institutions, one of which would be a small centre doing advanced research in selected areas, along with extension activities and international collaboration. Such a research centre should have PhD students, but should not be large. It should be located close to an existing institution of excellence and be able to use some of the facilities available in such an institution. After a few weeks, I received a call from Mani Shankar Aiyer, saying that the proposal had been approved and that I should proceed to prepare a proper description of the institute. With the help of some of my colleagues in the Science Advisory Council to the Prime Minister, details regarding a new scientific research centre to be established in the name of Pandit Nehru were provided. Soon the centre became a reality and the question was where it was going to be located arose. There were offers to locate it in Allahabad, Pune, etc., but eventually I was convinced by many colleagues that it should be in Bangalore. Rajiv Gandhi was particular that the new centre should be in Bangalore, close to the Indian Institute of Science.

I was asked to plan the centre and take the necessary steps to establish it. I was director of IISc and had to take up additional work for the new centre. It was named Jawaharlal Nehru Centre for Advanced Scientific Research (JNCASR). It was to be funded by the Department of Science and Technology of the Government of India. Fortunately, the Karnataka government gave nearly 20 acres of land free of charge in north Bangalore. With a grant of Rs 4 crore to build a campus, I requested Charles Correa to plan the new campus. In addition to the main building in the campus, we built a few residential quarters along with a student hostel. The Indian Institute of Science allowed us to have some space on its campus to build an office along with a house for the President of the centre and some guest accommodation. A few quarters for the faculty were to be built in the housing colony outside the campus. Academic activities of the centre were started right away using a small office at the Indian Institute of Science. These activities included summer training for undergraduates and mini symposia on chosen topics. The summer fellowship programme for undergraduates became a major success and continues even today. JNCASR was the first institution to initiate such a programme for undergraduates and many other institutions, including other science academies in the country, have instituted similar fellowship programmes.

During my Presidentship of INSA (1985–89), I tried to initiate some new activities. Status reports on several aspects of science and technology were brought out. Plans were made for a new building and the foundation stone for the building was laid by the then President of India, Zail Singh. I worked closely with H.Y. Mohanram and S.K. Joshi, who were scientific secretaries of the academy.

In IUPAC, there was great need to restructure the organization so that it dealt with the modern trends in chemistry. Restructuring and the associated changes in the various committees of IUPAC were initiated. I also began a series of monographs on important topics with the IUPAC sponsorship. I myself edited two IUPAC monographs on advanced materials and superconductivity. When I gave the presidential address of IUPAC in 1987, in Boston, I demonstrated the working of a high temperature superconductor. This was new for IUPAC.

High temperature superconductivity gave a tremendous boost for research in physical sciences. Thanks to Prime Minister Rajiv Gandhi, we could set up a special fund to support superconductivity research in educational institutions and research laboratories. A large community of scientists started work in this area. In my laboratory, we could make many new types of superconductors and also contribute to the studies of these materials. Many national and international conferences were held in India on high temperature superconductivity. I spent around nine weeks at Grenoble during the summer of 1990 as a visiting professor at Universite Joseph Fourier and worked with colleagues in Institut Laue–Langevin (ILL) on neutron diffraction studies of cuprate superconductors. I also worked closely with Benoy Chakraverty at LEPES, a laboratory founded by Louis Néel.

The Science Advisory Council to the PM initiated several new efforts, one of which was the establishment of TIFAC, Technology Development Fund, and a centre for developing hardware for high-level computing. One of the important functions of the council was to monitor technology missions in selected areas of national importance. The missions included literacy, malaria eradication, clean water, oil seeds, etc. The council also helped the Prime Minister arrive at conclusions or decisions on a variety of questions, such as the Antarctic expedition and cryogenic engine for rockets. I worked closely with Professor M.G.K. Menon on some of the issues.

**Rajiv Gandhi with members of the Science Advisory
Council to the Prime Minister (1988)**

Professor Menon was then member (science) in the Planning
Commission. The various reports of the council were published
as a book by Tata McGraw-Hill called *Perspectives in Science
and Technology*.

I was President of the Indian Science Congress in 1988, and
the annual session was held in Pune and inaugurated by Rajiv
Gandhi. I tried to modify the agenda of the congress by nearly
eliminating the presentation of short papers, by devoting most
sessions to lectures by experts, hoping that the they would be of
greater benefit to students and teachers. Rajiv Gandhi took the
trouble to be in a meeting as he thought that it was purposeful.
I cannot forget how he came to Kolkata from Delhi just for a
couple of hours to inaugurate the centenary of Professor Raman.
I was chairman of the national committee and had to delicately
plan the meeting to satisfy the demands of various groups.

It was enjoyable to work with a Prime Minister who was
enthusiastic and was always available to the council. I used to
see him often early in the morning at his house or later in his

Science Congress: With Rajiv Gandhi (1988)

parliament office. When it came to science, he had all the time in the world. He was keen to increase investment in science and technology to 2 per cent of the GDP. He was prepared to take tough stands and make quick decisions. He would ask ministers and bureaucrats to hasten things or bend the rules to get things moving. Rajiv Gandhi was mistakenly thought to be arrogant by

Rajiv Gandhi presenting the G.M. Modi Award to Rao. Professor M.G.K. Menon can be seen

some people. The problem was that he just could not stand second-rate work. He once got angry with a secretary to the government, who could not read what he had written. The transparencies prepared by him for projection were horrid. This did not mean that Gandhi punished him. He was a kind man and would excuse people for their mistakes easily. Even after he lost the elections, he would ask me to meet him to discuss what kind of structures should be set up in the government when he came back as Prime Minister.

When Chandra Shekhar became Prime Minister with the support of Rajiv Gandhi, I was asked to be member (science) in the planning commission. I was abroad when this suggestion was made by Gandhi. When I came back, he asked me to take up this task at least for a short while and prepare a science and technology plan. Within a few months, we prepared a progressive science and technology plan, but unfortunately, it did not see the light of day. Chandra Shekhar resigned after a few months and there was a new government. During the ensuing elections, Rajiv Gandhi was assassinated. When I was informed about this by phone, I cried like never before.

Chandra Shekhar's resignation from Prime Ministership was dramatic. There was a meeting of the CSIR society that morning, to be chaired by the Prime Minister. The meeting was held in one of the rooms in the Parliament and I was present as a member of the society. Soon after arriving in the room, Chandra Shekhar said that he had something urgent to attend to and asked me to chair the meeting. After an hour, he returned, by which time most of the items on the agenda had been discussed. He made closing remarks and the meeting ended. When we went out of the room, there were many MPs and others in the corridors. I learnt then that the PM had just submitted his resignation to the President of India.

Chandra Shekar

The government that followed, under P.V. Narasimha Rao, had no science advisory mechanism, though I was told once or twice by the Prime Minister that I should chair the Science Advisory Committee to the Cabinet (SACC). When I.K. Gujral became PM, I was asked to be chairman of SACC. Even in the short tenure of this government, we could establish the Fund for Infrastructure in Science and Technology (FIST) programme under DST.

I.K. Gujral

In 1988, my mother, my first teacher, passed away after an accident at home. When we looked in her almirahs and cupboards, we found nothing. She had given away whatever she had. Her generosity was remarkable. My father-in-law, Narayana Swamy, died in his sleep in 1989. He was a gentle soul and always cheerful.

Although fullerenes had been identified in mass spectra in 1985, they were successfully prepared in a laboratory in Germany only in 1990. I thought that we could do some work on fullerenes and started making the same in the laboratory. I found out that I could not do much research on the chemistry of fullerenes since we had no mass spectrometer and related facilities. It became a competitive area soon. I failed to get some of the organic chemists in the country interested in working on fullerenes. I, therefore, decided to work on phase transitions induced by temperature and pressure making use of C_{60} and C_{70} prepared by us. We carried out work on phase transitions, spectroscopy and photophysical properties in addition to molecular interactions involving C_{60} and C_{70}. Ajay Sood collaborated with me in some of the studies.

Carbon nanotubes were discovered in 1991, and we started making carbon nanotubes and contributed to many aspects of research on carbon nanotubes. I decided not to stick to nanotubes

With Pope John Paul II at the Pontifical Academy of Sciences (1990)

and fullerenes of carbon alone but also of inorganic layered materials. We started making nanotubes of metal sulfides and other inorganic layered compounds. This was a great success and resulted in major contributions from the laboratory.

In 1990, I was elected as member of the US National Academy of Sciences (NAS) and as well as of the Pontifical Academy of Sciences (PAS). PAS meetings are interesting and give members an opportunity to meet the Pope. It should be noted that based on the recommendations of PAS, Pope John Paul II exonerated Galileo and apologized on behalf of the church for having excommunicated him.

Between 1989–91, I was President of the Indian Academy of Sciences. M.G.K. Menon, Satish Dhawan, S. Varadarajan,

Sir C.V. Raman, founding President of the academy

Presidents of the Indian Academy of Sciences

S. Ramaseshan, Obaid Siddiqui and T. Sadasivan had preceded me as Presidents of the academy. Roddam Narasimha succeeded me. It felt great to serve this academy that I loved so. How can I ever possibly forget Professor Raman?

For several years, I was on the bureau and executive committee of ICSU. While it gave me much experience, I was not satisfied with the type of activities the organization dealt with.

PAS is different from other academies in many ways. One meets people from different backgrounds. Some of the people I have enjoyed meeting in this academy are Charlie Townes, Sir John Eccles, Cardinal Marini and the coordinator of the academy, Bishop Sorondo. I am tempted to briefly narrate how I enjoyed listening to Joseph Murray, a great surgeon from the US, who received the Nobel Prize for physiology and medicine for his work on organ transplant surgery. The stories he often narrated were very moving. He used to describe the pioneering advances made in surgical techniques to correct deformities in the head and face caused by accidents or birth defects. One of Dr Murray's patients was a twenty-two-year-old aviator named

Charles Woods. During war-time runs in the far east, his plane caught fire and, in his attempt to escape, he was burnt beyond recognition. His nose, eyelids and ears were obliterated, and by the time Woods reached the US, he was barely alive. Over the next two years, Dr Murray and his team performed twenty-four operations designed to build their brave patient a new face. Woods survived and went on to become highly successful, and to fly again.

Richard Herrick, whose kidneys were failing, was saved when his brother, Ronald, agreed to donate one of his healthy kidneys and thus made possible the first successful kidney transplant in the world, performed by Dr Murray. Throughout his career, Dr Murray travelled to other countries to help clinicians deal with patients suffering from craniofacial and other deformities, such as patients in India whose hands had been disfigured by leprosy but who, after surgery, were able to use their hands to work as artisans and become self-sufficient. Dr Murray talked of Ray McMillan, who was abandoned by his mother because of a birth deformity and was placed in a mental health institution. When he was twenty-one, Ray's grandmother rescued him from this senseless incarceration. Dr Murray was able to correct Ray's facial deformity and he encouraged the boy to write what was in his heart. To read what this 'mentally disabled' boy had to say, and about the joy he found in life, was to witness the transforming effect of Dr Murray's care and skill in restoring not just bodies but souls as well.

In 1983, a major world academy, the Third World Academy of Sciences (TWAS) was started at the initiative of Professor Abdus Salam, with its office in Trieste, and I was one of the founding members. The purpose of this academy was to help scientists in the developing world do science and also get due importance for science in the developing world. This was based on the faith that every country, however poor and however small, requires a scientific

**At the general conference of TWAS, New Delhi,
with Professor M.M. Joshi (2003)**

base for progress. I took much interest in the activities of TWAS from 1986 onwards and prepared its statutes. I worked closely with Professor Salam in the early years. When Professor Salam became ill after some years, I was named Vice President of the academy (between 1997–2000). I promised him that I would protect the interests of the academy.

In 1992–93, I was elected member of the American Philosophical Society (APS), the oldest academy in the US, founded by Benjamin Franklin. I received letters from many friends in the US congratulating me on the hat-trick. This was because I was

**TWAS meeting in China
(with President Hu)**

AMERICAN PHILOSOPHICAL SOCIETY
HELD AT PHILADELPHIA FOR PROMOTING USEFUL KNOWLEDGE
104 South Fifth Street, Philadelphia, PA 19106-3387
Phone: (215) 440-3400 • FAX (215) 440-3436

Herman H. Goldstine — Executive Officer

April 21, 1995

Dr. C. N. R. Rao
Indian Institute of Science
Solidstate and Structural Chemistry Unit
Bangalore 560012
INDIA

Dear Dr. Rao:

This notification of your election today as a foreign member of the American Philosophical Society, on a form used by the Society for a very long time, may require an explanation.

The American Philosophical Society was founded by Benjamin Franklin in 1743, in emulation of the Royal Society and the Dublin Society of Arts. The APS has played an important role in promoting science and learning in the United States and recognizing scholarly achievement everywhere. Lavoisier, Darwin, Einstein, Pasteur, Lord Rutherford, and Mme. Curie have preceded you among our foreign members.

Under separate cover you will receive a copy of the 1993 Year Book of the Society, which contains the Laws and List of Members, a short history, and a full account of the Society's recent activities. The 1994 edition will not be published until this summer.

As you will see, today the Society supports research through a program of grants, publishes scholarly books and journals, maintains a library singularly rich in material in the history of science and technology. Its meetings express in the twentieth century the universal spirit of the eighteenth century "Age of Reason" with the presentation of papers on such topics as Jefferson and the ambiguities of liberty, superconductivity, cancer control, the history of computers, rising health care costs, Shakespeare, and education and our divided knowledge.

The next Autumn General Meeting will be held on November 2 - 3, 1995. The next Annual General Meeting occurs on April 25-27, 1996. Both meetings occur in the newly renovated Benjamin Franklin Hall. The building opened in 1993 when the Society celebrated the 250th Anniversary of its founding and of the birth of its third President, Thomas Jefferson. Meanwhile, after we receive your acceptance of election, your certificate of membership will be sent to you.

Sincerely,

separate cover

Philosophical Hall • 1789   Benjamin Franklin Hall • 1855   Library Hall • 1959

**Receiving the UNESCO Albert Einstein Gold Medal from
Dr Federico Mayor, director general (1996)**

already a member of the American Academy of Arts and Science (1986) and of the National Academy of Sciences (1990). APS is an amazing academy and organizes interesting annual meetings.

Another major area of research that we started working on during 1993–94 related to colossal magnetoresistance (CMR) in oxide perovskites of manganese. They not only exhibit colossal magnetoresistance, but also important phenomena such as charge and orbital ordering as well as electronic phase separation. The phase separation results in having different types of electrons exhibiting different properties and phenomena in the same crystal. I got heavily involved in research in this area and published a large number of papers, many along with Arup Raychaudhuri.

The period between 1989–94 was a difficult one because of my various commitments and also the urge to do research in new areas. It was during this period that I was involved in founding two important professional societies—the Materials Research Society of India (MRSI) and the Chemical Research Society of India (CRSI). Both these have since emerged to become active societies providing the much-needed boost to materials research and chemical research in the country.

I worked for four to six weeks at the University of Cardiff during the summers of 1992–94, collaborating with Wyn Roberts on surface science. I had a research grant from Unilever to support my research there. I was quite exhausted by the end of 1994. It became clear that if one wants to accomplish major research results and work in competitive areas, one should not be doing administration. Fortunately, in 1994, at the age of sixty, I could give up the directorship of the Indian Institute of Science. I was pressurized by Dr Ramanna and others to continue as director for one more term up to the age of sixty-five, but I was absolutely certain that there was no way I was going to take on this responsibility any longer. Even though I retired from the Indian Institute of Science, on 31 July 1994, I had to build JNCASR. I started working partly in

JNCASR, which had a campus by 1994. I was asked to be President of JNCASR for the five year period between 1994–99.

During my last year as director of IISc, the employees of the institute informed me that they were urging the Bangalore Corporation to name the circle in front of the institute after me. I requested them not to do this. They were competing with the employees of an industry opposite the

With Dr Raja Ramanna (chairman of IISc council)

institute about naming the circle. I expressed my discomfort to Raja Ramanna. I forgot about the matter until one day, the employees came to my office and requested me to go to the front of the institute to witness the naming of the circle after me. The mayor of the city was waiting for me. I found it most embarrassing and yet had to be with the mayor for the naming ceremony. This is an example of a situation that one faces beyond one's control. The circle no longer exists and an underpass has since been constructed at the location.

When I reached the age of sixty, IISc made me a lifetime honorary professor and conferred its fellowship on me. My colleagues at IISc organized a major function and several international journals, including the *Journal of Physical Chemistry*, brought out special issues. *World Scientific* brought out a volume titled *Solid State Chemistry* containing selected papers of mine. At home, my wife and I got remarried following an age-old custom.

When I retired from IISc in 1994, I found my health was not as good as it used to be. I went back to my friend Dr K.G. Nair in Bombay. After various tests, he said that I should go through a redo heart surgery. I underwent the surgery in February 1995 at Breach Candy Hospital with Dr Panda doing the operation. I came out of it successfully and started working within three weeks. The year 1995 was an important one in our family. My granddaughter Suguna was born on 6 December.

During this period, I faced some unfortunate experiences. One or two scientists accused me of being interested in continuing at IISc forever. They had an objection to JNCASR being given space at IISc. The idea of establishing a new centre in the name of Nehru included the requirement that some facilities and space be provided by an existing institution, which had been agreed to by the council and trustees of IISc. (Some of the facilities set up by JNCASR are now being used by IISc as well. Some of the old buildings used by JNCASR after renovation have since been returned to IISc.) The criticism that I was planning to be at IISc as the director or as the chairman of the council was completely unfounded because I was anxious to give up the directorship. I had not been interested in being the director of IISc in the first place, as I had no interest in running it or any other institution.

Also, after JNCASR started functioning in its new campus, some people who visited it made negative comments stating that JNCASR was taking credit for some work its honorary professors

**Jawaharlal Nehru Centre for
Advanced Scientific Research (JNACSR), Bangalore**

had done by listing some of their papers in its reports. JNCASR had several honorary professors and some of them had been given financial grants from the centre for research in the initial days of its establishment. They were supposed to give a list of papers arising from the work supported by JNCASR. It is possible that some of them included other papers as well in the list. It was unfortunate that some scientists wrote about such matters without knowing the facts and without appreciating the spirit of dedication with which the centre was being built. Accusing colleagues of insincerity or lack of integrity not only discourages them, but often stops them from functioning. I have refused to yield to such negative comments by people, although they have hurt me deeply.

By 1994, JNCASR had a fine campus in the north of Bangalore (near Jakkur). K.R. Narayanan, Vice President of India, inaugurated the campus. By 1994, some experimental work was started in JNCASR and the first faculty member (Dr K.S. Narayan) joined. We soon established some facilities including an electron microscope, X-ray diffractometers, photoemission spectrometer and so on. During the 1994–99 period, efforts were made to build different units devoted to the chemistry and physics of materials,

K.R. Narayanan dedicating JNCASR to the nation

engineering mechanics, and molecular biology and genetics and evolutionary biology. I took on a few students for PhD work at JNCASR. By 1999, the various units had taken shape, with faculty members appointed and research publications flowing from the centre. JNCASR had established itself as a bona fide research institute by 1999. In order to give PhD degrees to students, the centre had come to an agreement with Manipal Academy in the initial years. After 1999, the centre became a deemed university and could give its own degrees. I started publishing research papers from the centre with PhD students as well as integrated PhD students (doing MS and PhD degrees). I initiated some research on organic supramolecular structures and on nanomaterials.

The centre started an educational technology unit, which has been bringing out science education material for schoolchildren in the form of booklets, CDs and so on. In 1995, a book of mine on transition metal oxides (Wiley-VCH) came out. The book was co-authored by Bernard Raveau of France. Its second edition was printed shortly after. A book on colossal magnetoresistance edited by me and Raveau also came out around this time.

A Festschrift devoted to metal-insulator transitions to celebrate Professor Mott's ninetieth birthday was edited by Edwards and me, and published by Taylor and Francis (1995). In the following year, we organized a discussion meeting in the Royal Society on metal-insulator transitions. Professor Mott wanted to attend the meeting and make a few comments. Unfortunately, he

University of California, Santa Barbara (UCSB)

passed away before the meeting took place and we dedicated the proceeding to his memory.

From the summer of 1994, I started spending six to eight weeks in the University of California, Santa Barbara, which had appointed me as a distinguished visiting professor in the department of materials. The department was outstanding, with a large number of national academy members in the faculty. During my stay at Santa Barbara, I helped initiate a major programme on metal-organic framework structures. I collaborated with Tony Cheetham, who was director of the materials research laboratory at Santa Barbara. UC Santa Barbara treated me well and living in Santa Barbara was like living in a dream world.

In 1997, my father passed away at the age of ninety-four. He was a great supporter of mine, especially when I was young. He had great faith in me. He remained a highly disciplined person throughout his life, reading every newspaper he could get hold of. He maintained an interest in cricket, and paid lots of attention to his health.

I was invited to be Linnett Professor of physical chemistry in the University of Cambridge in 1998. I really enjoyed my stay in Cambridge for those six weeks, living in an apartment in Sidney Sussex College, located in the heart of Cambridge. During my stay, I planned to edit a book titled *Supramolecular Organization and Materials Design* along with Bill Jones. The book was later published by Cambridge University Press.

In 1999, I retired from the Presidentship of JNCASR. JNCASR allowed me to take students as long as I wanted and conferred its fellowship on me. I was made Linus Pauling Research Professor from the endowment received from Reliance Industries. By then, around eighty people working with me had received PhD degrees. I wanted to spend the next few years working in the laboratory working with PhD students and postdoctoral fellows.

In closing

The years 1984–99 was a very busy period in my life. While I did as much as I could in research and also contributed to other spheres of activity that required my involvement, I had learnt some important lessons—the most important one being that one should not take up a full-time administrative position if one wants to do outstanding research. The two are mutually exclusive. The second lesson has to do with time. Time is precious and one has to learn how best to use it.

'What is the longest and swiftest thing in the world; the swiftest and the most slow; the most divisible and most extended; the least valued and the most regretted; without which nothing can be done and which devours all that is small and gives life and spirit to everything that is great. It is what the creator thought of such value as never to bestow on us mortals the minutest part of it. It is time'—Michael Faraday

6. WORKING FOR PLEASURE: LIFE AFTER SIXTY

After relinquishing the directorship of the Indian Institute of Science in 1994, I made considerable effort to build the new Jawaharlal Nehru Centre for Advanced Scientific Research. It was enjoyable in many ways and the centre initiated work in units dealing with materials, biology and engineering mechanics by 1999. After relinquishing the Presidentship of JNCASR in 1999, my administrative days came to an end. It was a wonderful feeling to be free to do whatever I wanted. A.B. Vajpayee's government did not have any science advisory mechanism. While drafting the new technology policy of the government, Dr M.M. Joshi, the science minister, did not involve me in any way. This added to my freedom. I felt that I could do research in a relaxed atmosphere. Around this time, nanoscience was being recognized as an important area all over the world. I was doing research on nanomaterials even in the 1980s, well before the subject was recognized as important. I used to work on small metal clusters (nano particles), their electronic properties as well as their assemblies. I had discovered how metal nano particles became non-metallic when they were very small with diameters of around one nanometer (containing 100–200 atoms).

After my sixtieth year in 1994, my research performance showed considerable improvement. There was a definite increase in citations, possibly because I had fewer worries. Without any administrative responsibility, I hoped to do much better after 1999. I continued to research various aspects of carbon nanotubes, and on the synthesis and characterization of inorganic nanowires. Some work on carbon nanohorns was also pursued. We could make nanotubes of layered materials, such as MoS_2, WS_2, $MoSe_2$, GaS and so on. I had a few students synthesizing other types of materials. In metal-organic frameworks (MOFs), we could obtain several interesting results not only in synthesizing novel materials, but in understanding the mechanism of formation of these materials. A new structural classification of MOFs was proposed. The Royal Society awarded me the Hughes Medal in the year 2000. The Hughes Medal is awarded in recognition of an outstanding discovery in physical sciences. This award was unexpected and acted as a great morale booster.

The twenty-first century saw the discovery of graphene. This created the biggest sensation in recent times. One could not believe that a two-dimensional one-atom thick carbon sheet possessing such unusual electronic properties had been discovered. Graphene is gapless, with ballistic conduction of electrons. We started working on various aspects of graphene by synthesizing it by chemical means, instead of the scotch tape method of physicists. Two important articles that I published in *Angewandte Chemie* and the *Journal of Materials Chemistry* have been cited extensively.

I had a rather unpleasant experience during this time. A student was jointly working with me and another professor from IISc, and did some good work on the use of graphene for radiation detection. The paper was accepted for publication in a leading journal *(Advanced Materials)*. Soon after, we found that a few sentences in the introduction had been reproduced by the student from a paper in the literature. Unfortunately, I had checked only the results section since the paper had already been seen by the co-guide. I immediately wrote to the editor withdrawing the paper. The editor wrote back that we should not withdraw the paper since it was very good and that she would insert an appropriate apology in the journal. The apology note published later was not worded properly and a few people accused us of plagiarism. As far as I was concerned, the editor was the final judge in the matter and she had fully endorsed the paper's publication because of its high quality. Things died down after the matter was adequately

explained. Many of my friends abroad were shocked by the
unwarranted reactions in India on a matter of routine occurrence.

In 2004, a new government was formed in India with Dr
Manmohan Singh as Prime Minister. Singh asked me to be chairman
of the new Science Advisory Council to the Prime Minister. It had
a small membership, with the heads of scientific agencies as invitees.
Some of the members were Roddam Narasimha, P. Rama Rao, G.
Mehta, M.M. Sharma, Kiran Karnik, R.A. Mashelkar, Baldev Raj, K.
Vijay Raghavan, A. Jhunjhunwala, T.V. Ramakrishnan, P. Balaram,
A.K. Sood, S.E. Hasnain, V.K. Singh, B. Sinha and B.K. Thelma. I
took up this responsibility and worked as chairman of the council for
a period of ten years, until 2014. The secretary of the Department of
Science and Technology was member-secretary of the council, and
all the heads of science agencies and departments were observers. V.S.
Ramamurthy occupied the position of secretary of the council for the
initial four years and T. Ramasami for the rest of the duration of the
council.

**With Prime Minister Manmohan Singh and
Kapil Sibal at the release of the SAC-PM report**

Prime Minister Manmohan Singh with members of the Science Advisory Council

Remembering how the discovery of fullerenes and nanotubes of carbon prompted research on the inorganic analogues of these zero- and one-dimensional materials, I decided to work on other two-dimensional layered materials (inorganic analogues of graphene) in 2004–05. We worked on single and few-layer species of these materials, particularly of molybdenum disulfide (MoS_2) and other dichalcogenides, and published some of the first papers on the subject. We followed this by work on boron nitride. Starting early on two-dimensional inorganic materials proved to be fortunate for me. This area has become highly popular in the last five to six years and MoS_2 is one of the novel materials with the most unusual properties. In some ways, these materials are more interesting than graphene since they have band gaps.

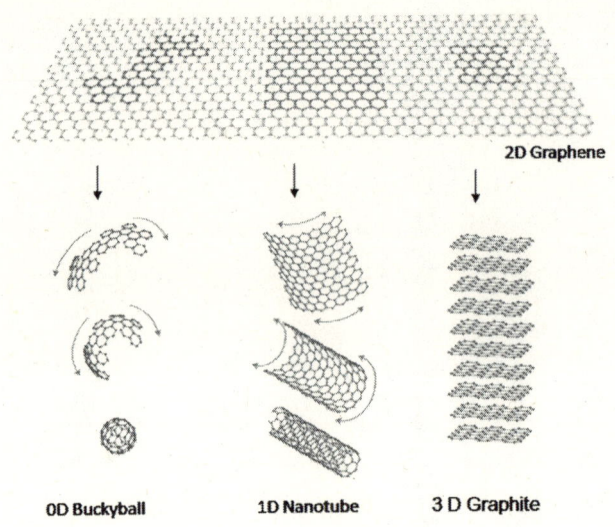

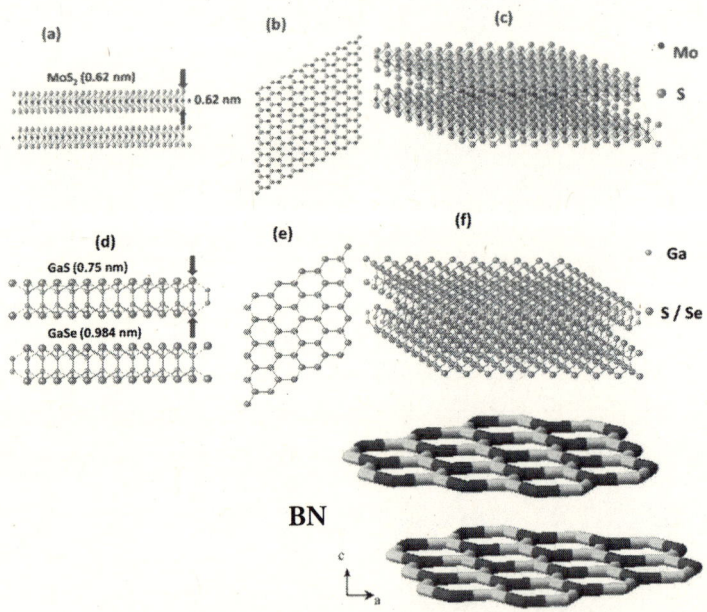

Graphene and inorganic graphene analogues

Rao receiving the India Science Prize (2004)

In 2004, I received the first India Science Prize, the highest recognition in science by the Government of India. The award was presented by Prime Minister Singh at the Indian Science Congress. That year, I also received the million-dollar Dan David Prize from Israel (known as the 'Israeli Nobel') along with Robert Langer of MIT and George Whitesides of Harvard. My wife and I went to Tel Aviv to receive the prize and took the opportunity to travel around Israel.

In 2000, I became the President of the Third World Academy of Sciences (TWAS), now called The World Academy of Sciences and worked in this capacity for six years. During this period, I worked closely with Mohamed Hassan (from Sudan) and Jacob Palis (from Brazil), and could get to know scientists from various countries, particularly Africa. It is fantastic to be with scientist friends from Africa and other developing countries. They all have every hope and aspiration, but face major hurdles in their progress. Working for TWAS was a pleasure and I could do many things as its President, besides electing new fellows and organizing meetings.

**With Prof. Whitesides and Prof. Langer at the Dan David Prize
function, Tel Aviv (2005)**

**Rao receiving the 'Karnataka Ratna' from the then chief
minister, S.M. Krishna (2001)**

A special research grant system for the best scientists from the
least developed countries was established, besides the spare parts
and mini grant programmes. Regional chapters of TWAS were

Rao at the ceremony to receive the Chevalier de La Legion d'Honneur award from France (2005)

established in different parts of the world. In the last few years, TWAS has emerged as the voice of science in the developing world. I have worked for TWAS for over eighteen years in various capacities, and have chaired committees for selecting fellows and awarding prizes. I continue to be part of TWAS, and chair the

With Jacob Palis and Mohamed H.A. Hassan (TWAS)

prize committee. In the history of TWAS published recently, I was flattered to see that Mohamed Hassan, who served as executive director of TWAS since its inception, has given me credit for making TWAS the voice for science in the south.

Of the many activities initiated at JNCASR, an important one has to do with science education and outreach programmes. In order to support science education, my wife and I founded the C.N.R. Rao Education Foundation in 2005. For this purpose, we used nearly 50 per cent of the money earned by me through various awards. The foundation supports activities in science education in various parts of India. In particular, it supports science education

The science outreach programme

programmes in the Himalayas. Every year, we go to Uttarakhand, where we conduct workshops for schoolchildren and teachers. We have been able to build facilities for holding workshops in Gangolihat (close to Pithoragarh). We have also conducted education programmes in Kerala for the last thirteen years. As far as possible, the programmes are held in interior India and remote areas. We have covered several places in Assam, Bengal, Kerala, Karnataka and a few other states. The foundation also awards a prize to the best research scientist from a least developed country through TWAS.

A Hall of Science was built in the JNCASR campus where we conduct programmes for schoolchildren regularly. My wife and I donated a major part of the cost of building through our foundation, and received some contribution from Mukesh Ambani of Reliance Industries. The Hall of Science is an active place where children come from all over to listen to lectures and do experiments. It houses an exposition on materials chemistry, the C.N.R. Rao archives and a viewing room with computers for the use of children. It conducts workshops for teachers and gives national awards for science teachers, and produces educational materials like books, CDs and DVDs. My wife and I have contributed

**Richard Zare and Bengt Nordan at the inauguration
of the Madan Mohan Malaviya Amphitheatre**

C.N.R. Rao Hall of Science, JNCASR

to these activities by direct participation, and have produced several books, CD's and other materials. The important books produced are *Learning Science* (in four parts, for children from classes eight to ten) authored by Indu and me, *Understanding Chemistry, Nanoworld and Nano Carbons* authored by me. They have been translated into several Indian languages and some

Inauguration of the C.N.R. Rao Hall of Science and dedication of ICMS to the nation by Prime Minister Manmohan Singh (3 December 2008)

The International Centre for Materials Science (ICMS)

have been published by the National Book Trust. The purpose of the outreach programmes is not only to help provide proper science education to children but also to pique their imagination and promote scientific temper. The Hall of Science was inaugurated by Prime Minister Manmohan Singh in 2005. At that time, he also inaugurated the new building to house the International Centre for Materials Science, funded by the Department of Science and Technology.

The International Centre for Materials Science (ICMS) carries out research in advanced materials and provides opportunities for young scientists and teachers from different institutions, including those from developing countries in Africa and elsewhere, to visit for short periods. Within India, the centre organizes several meetings and symposia. It has established excellent facilities for electron microscopy, molecular beam epitaxy, nanoscience and nano fabrication, crystallography, high pressure synthesis, photoemission spectroscopy and scanning probe microscopies, and so on. The laboratory building of the centre is truly attractive and has been admired by many a visitor. In the last six–seven years, a new unit devoted to chemistry, called the New Chemistry Unit (NCU), has been set up. The faculty of this unit works in areas close to materials or biology. The broad areas of materials research covered at JNCASR today include energy materials, magnetic and electronic materials, nanomaterials, polymers, supramolecular materials, sorption materials, as well as theory and computation.

In 2008, I was asked to be a member of an advisory committee to set up an advanced centre for materials in Ras al-Khaimah, one of the emirates in the UAE. The committee was chaired by Tony Cheetham, and included Millie Dresselhaus, Richard Friend and Mike Klein as the other members. We met a few times and planned a laboratory and related requirements with a budget estimate of around $40 million. With an interest to see a working materials research centre, Sheikh Saud came to Bangalore aboard his own plane. He seemed quite impressed by JNCASR. When it

Sheikh Saqr Laboratory

was eventually decided not to set up a materials research centre in RAK due to a paucity of funds, Sheikh Saud decided to fund materials research elsewhere and picked JNCASR for this purpose. JNCASR is receiving a sum close to $4 million over a period of six years, including some funds for part of a laboratory building. Sheikh Saqr Laboratory (SSL) was thus set up as part of the ICMS, part of the new chemistry building named after Sheikh Saqr (the late father of Sheikh Saud). The annual running budget of SSL is $600,000. This is used to provide research support for faculty members and much of it is used for research equipment. SSL also supports research personnel, and gives fellowships to faculty members and research students. It organizes topical

Inauguration of the Sheikh Saqr Laboratory by Sheikh Saud

workshops on materials and chemistry. SSL has been of great benefit to us all. The New Chemistry Unit (NCU) in the centre is growing along with SSL. The new building of NCU, which houses SSL, was inaugurated by Sheikh Saud in 2013. Sheikh Saud's donation is without any conditions and has no strings attached. I have not seen the type of generosity and interest shown by Sheikh Saud in any of the Indian industrialists.

In 2009, the Indian Institute of Science celebrated its centenary. As part of the celebrations, I was asked to give the first centenary lecture and the institute published a volume containing selected research papers of mine through World Scientific Publishers. The centenary celebrations coincided with

Conferment of the 'Order of Friendship' by the Russian Deputy Prime Minister (2009)

my seventy-fifth birthday. Since I felt I was getting old, I thought that I would write an autobiography. I wrote the book *Climbing the Limitless Ladder* and it was published by World Scientific.

In 2009, the Royal Society awarded me the prestigious Royal Medal (Queen's Medal). This was the second major recognition from the Royal Society for my work. The Royal Society has always been very encouraging of my academic endeavours. In 2010, the Royal Society celebrated its 350th anniversary and I was asked to respond on behalf of all the fellows. The grand function was held at the Royal Festival Hall in London and the Queen inaugurated the meeting. She had come with all the members of the royal family for the occasion. I felt honoured to deliver the address on such an occasion, attended by the heads of all academies and nearly 2000 fellows and other scientists.

Rao receiving The Royal Medal from The Royal Society (2009)

**Rao responding on behalf of the Fellowship at the 350th
Anniversary of the Royal Society, London**

Between 2004–10, several books of mine were published. Three books related to nanomaterials and nanochemistry, as well as graphene, were all published by Wiley-VCH. A book on nanotubes and nanowires was published by the Royal Society of Chemistry, London.

The German Chemical Society awarded me the August Wilhelm von Hofmann Medal in 2010. This medal is an international award given by the German Chemical Society to a leading chemist in the world. In 2011, I received the Illy Trieste Science Prize for materials research. Over the years, I have given plenary or keynote lectures at innumerable international conferences related to the chemistry of materials. I have also delivered named lectures in several universities around the world. I must especially mention my visit to the Massachusetts Institute of Technology to deliver the A.D. Little lecture in the chemistry department in 2010. Going around MIT, I did feel like I was going through the corridors of probably one of the best institutions in the world. MIT represents the very best in science and technology, and has the necessary

August Wilhelm von Hofmann Medal. Prof. Martin Jansen is standing on the right of Rao

Ernesto Illy Trieste Science Prize

cross-disciplinary spirit that promotes modern education and research, besides innovation. A large number of start-up companies and industries have come up based on the research and development carried out by MIT faculty.

In the last few years, I have received honorary doctorate degrees from several universities and some of the convocations have been memorable. In Oxford, the ceremony was conducted in Latin and the other recipient of the honorary doctorate was Jimmy Carter, former President of the US. At Uppsala, the rector gives a gold ring to the recipient and places a crown of leaves on their head. As he does so, a cannon is fired. At Uppsala, Nelson Mandela was a recipient of an honorary doctorate degree at the same convocation as I. At St Andrews, I received the honorary doctorate with Hillary Clinton, the Archbishop of Canterbury and Jane Goodall. These were some of the occasions when I became a celebrity for a moment.

Since 2010, I have been exploring some new areas of research. One of these relates to the splitting of water by artificial photosynthesis. We have found success both in producing oxygen by the oxidation of water and hydrogen by the reduction of water using semiconductor heterostructures and dye sensitization. We have also carried out studies on solar thermochemical decomposition of CO_2 and H_2O to yield CO and H_2, a viable method to produce syn gas. We have found a low-cost non-metallic catalyst to replace platinum for the electrolysis of water. In designing and synthesizing new materials, a novel contribution has been to make covalently cross-linked nanomaterials obtained by the linking of graphene and carbon nanotubes with themselves or with other materials such as MoFs and boron nitride. Our work on graphene analogues of inorganic layered materials seems to have received considerable attention. I have been invited to write articles in major journals on this topic and we are continuing to work in this broad area.

In this context, one of the important contributions made by us is to prepare borocarbonitrides, $B_x C_y N_z$, over a wide range of compositions, comprising insulating boron nitride and conducting graphene. These materials have been made in different ways and their electronic structure, gas adsorption, applications in energy devices and catalytic properties have been investigated. This has been a rich area of research and has yielded many useful applications in energy technologies

**Receiving the Desikottama
Award at Shantiniketan**

Honoris Causa, Uppsala

**Signing the register at the
Sheldonian theatre, Oxford**

**Honoris Causa,
University of Kashmir**

Honoris Causa, St Andrews (2013).
Hillary Clinton can be seen near me

Cartoon of Rao used in
the Molecular Frontiers
Symposium (2013) in
South Korea

related to batteries, fuel cells and supercapacitors. Although I was still interested in MoFs, I decided to give up working in this area since it was getting over-crowded. The only thing that I am doing with MOFs is to make composites wherein they are covalently bonded to graphene and other 2D nanosheets. Since 2008, I have stopped going to UC Santa Barbara during the summer as Indu finds it difficult to travel the distance. Tony Cheetham, with whom I collaborated, left UCSB for Cambridge University around this time. UCSB has, however, continued to show me as a visiting professor of materials. In general, I am finding it difficult to go abroad for long periods now.

An entirely new class of materials involving aliovalent anion substitution in metal oxides and chalcogenides has been explored. An example of this is provided by ZnO wherein oxygen is substituted by N and F. $ZnO_{1-2x}N_xF_x$ has the same number of electrons as ZnO, but possesses entirely different properties. Thus, when x=0.15 or so, ZnO is orange coloured while ZnO is white in colour. CdS wherein S is substituted by P and Cl is another example. Such substitution brings down the band gap, but the

July 5th, 2011

**With students celebrating the diamond
jubilee of his research career**

fully substituted aliovalent anion based materials such as Zn_2NF or Cd_2PCl are semiconductors possessing interesting properties such as photochemical hydrogen evolution from water.

In July 2013, our dear grandson, Dr Kartik, M.D, got married to Avantika, a fine Rajasthani girl. Avantika is also a medical doctor, who specializes in pathology.

In November 2013, JNCASR celebrated its silver jubilee. It was a heart-warming experience to witness the growth and performance of the centre. I felt like a proud father (grandfather!). The Vice President of India, Hamid Ansari, was the chief guest at a special function organized by the centre. It was a pleasure to have an enlightened scholar and friend like him at the occasion.

The Science Advisory Council to the Prime Minister, of which I was chairman from 2004 to 2014, has contributed to various aspects of science and technology in the country. During this period, I worked closely with Prime Minister Manmohan Singh, who was always cordial and encouraging. He welcomed new ideas and initiatives, and did not turn down any recommendation of the council. Based on the recommendations of the council, five Indian Institutes of Science Education and Research (IISERs) were established. I had the pleasure of presenting the IISER concept to the Prime Minister and others, on behalf of the education minister. These institutes promise to be new fountains of scientific talent. J.C. Bose fellowships for senior scientists, Ramanujan fellowships for aspiring scientists and many other important opportunities have been created. Two new science departments of the Government of India were established based on the recommendation of the council—the Department of Earth and Ocean Sciences and the Department of Health Research.

It was time the country had a Department of Health Research that took care of medical research and the work of the Indian Council of Medical Research. There were efforts to have such a department in the Government of India for many years, but it was only during the tenure of this council that it became possible. There were many other things that the council could do, covering areas such as biotechnology, agriculture and so on. The council has brought out a volume titled *Science in India*, which presents the accomplishments of the Science Advisory Council, and discusses the problems and challenges facing India. It provides an action plan to make India a global leader in scientific research and innovation.

An important event in the last ten years has been the setting up of the national nanomission to promote research in nanoscience and technology, of which I have been chair. It has been a pleasure to create facilities and thematic units for nano research all over the country, and set up two institutes. This has enabled India to emerge as a leader in the subject and come third in the world in terms of publications. Many technologies have also been innovated. Considering that there was hardly any nano research in the country before the commission was set up, its performance is noteworthy. National and international conferences organized by the mission have added to the success of the programme. It would be worthwhile to consider supporting a few other important areas of science and technology in the mission mode just as the nanomission.

Since 2008, I have been chairing the Karnataka government's vision group on science and technology. The chief ministers have all been supportive and I have enjoyed contributing to the state. In particular, we have conducted teacher training (refresher) programmes at all levels, rewarded good teachers and researchers in state institutions and awarded research grants to individuals, besides providing infrastructure support to colleges. To my embarrassment, I have been accorded a cabinet rank by the state. I have also been involved in organizing the annual Bangalore Nano, which has gradually emerged to become a national event.

My wife and I had gone to Trivandrum for our annual visit to give lectures to school children and postgraduate students in November 2013. On our way back, when I was in the airport, I received a phone call from Dr Manmohan Singh, informing

Pranab Mukherjee, President of India, conferring the Bharat Ratna on Professor Rao

me that I had been awarded Bharat Ratna, the highest civilian honour of the country. It was difficult for me to believe that I was the recipient of such a great honour. The science and technology departments of the Government of India, including the research councils, organized a reception on 14 January 2014, after a felicitation meeting ceremony chaired by the minister of science. That evening, the Prime Minister hosted a select group of scientists for tea at his house. The function awarding the Bharat Ratna in the Rashtrapati Bhavan was grand. It was held on 4 February 2014 and President Pranab Mukherjee was gracious enough to have tea with me and my wife after the ceremony. The Prime Minister was also present. The ceremony was attended by the entire family and we had a nice lunch after the ceremony.

Rao and family at the Rashtrapati Bhavan

Rao at the felicitation function in the Prime Minister's residence

Rao being felicitated by the
Karnataka government

Felicitation at
Rohtang Pass
(held at 13,050 ft),
Himachal Pradesh

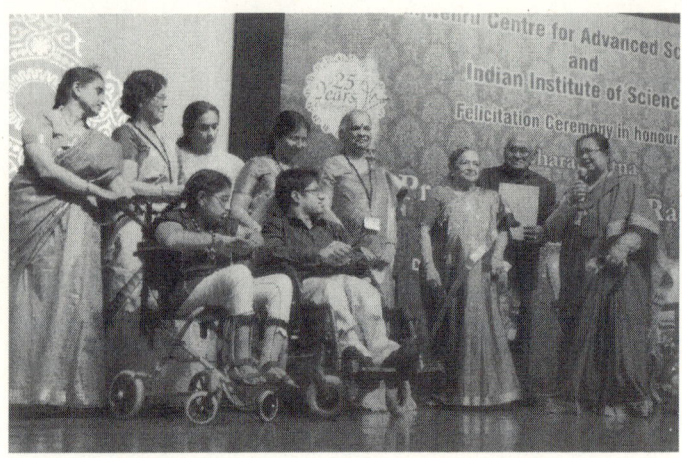

Professor and Mrs Rao with children from the spastic society, at the felicitation by JNCASR and IISc

In early 2014, I was elected foreign member of the Chinese Academy of Sciences, of which I was Einstein Professor in 2013. The Chinese Academy also awarded me its international medal.

International Award for Science, Chinese Academy of Sciences. With President Chunli Bai

When I turned eighty in 2014, World Scientific Publishers brought out a volume containing a selection of my papers on the chemistry of materials. The Angewandte Chemie family published a special issue of a journal on that occasion. There was also an international symposium held in Bangalore. The Royal Society and the Royal Society of Chemistry arranged a special symposium in London in June 2014 to celebrate my eightieth year. To mark the occasion, I visited Badrinath in the Himalayas with my wife and other members of our family.

The emperor of Japan awarded me the country's highest civilian honour, The Order of the Rising Sun, Gold and Silver Star in 2015. Since we could not go to Tokyo, a function was organized at the Japanese ambassador's house in Delhi. The year 2015 was also important in other ways. A book, *Lives and Times of Great Pioneers of Chemistry*, which I co-wrote with my wife, was published. Writing this book was a labour of love. It gave me a way to record

Order of the Rising Sun, Gold and Silver Star, Japan

my gratefulness to the founders of chemistry, considering that I have had a wonderful life doing chemical research for over sixty years. Writing this book had a profound effect on me. It humbled me and, at the same time, made me feel proud of being a chemist. A majority of the great pioneers of chemistry came from humble backgrounds and some of them were actually quite poor. They became great chemists because of their determination and hard work.

In 2015, my book *Essentials of Inorganic Materials Synthesis* was also published (Wiley, New York). In October that year, my wife and I became great-grandparents to a beautiful little girl child born to Kartik and Avantika.

I can say with pride and certainty that I have thoroughly enjoyed doing science, especially with young students. Engaging with them not only keeps me young, but also enables me to play some part in facilitating the emergence of young talent. What is interesting is that my research performance, which had shown a marked improvement after the age of sixty, flourished further after the age of seventy. The number of citations and the H-index both showed big jumps after I turned seventy. In the last few years, I have been cited around 6000 times each year and the total number of citations today is more than 88,000, while my H-index has reached 131. The i10 index (which gives the number of papers cited at least ten times) is around 1260. The total number of my research papers (after carrying out research for more than sixty years with a few hundred co-workers) has reached 1600. The number of books authored or edited by me is fifty. I never expected to touch such a record. I do not track the growth of these numbers, but the latest numbers have come as a pleasant surprise.

One can only keep doing good work, leaving the results, in terms of recognition and citations, in the hands of others.

What is particularly gratifying is that one's performance in research is not particularly related to one's age. I am reminded of a famous composition of Sant Kabir Das. I have come to the conclusion that everything is in one's mind. You are as old as you think you are. The real problem, however, is that others decide on things based on one's age. In India, an individual is made to retire from service at the age of sixty or sixty-five. Clearly, at sixty or sixty-five, one is quite energetic and capable of many more years of productive work. I feel that it should be possible for good scientists to continue working as long as they can (and want to), with or without remuneration. After I turned sixty, someone said to me, 'I believe you have retired.' I had to quote A.B. Vajpayee at that time: 'I am not even tired. How can I be retired?' A few weeks ago, I was surprised a major university in the US asked me to be an adjunct professor for a period of five years. I am sure they are aware my age.

I am facing a special type of problem. As I am getting older, I wonder whether I should take on new students for PhD work with me. Any student that I take on now will require five or six years to finish, and I would be in my late eighties by then. It is not easy to have a young person as a second guide, partly because the personal ambitions of the young person may not allow him or her to play such a role. Unfortunately, the quality of postdoctoral fellows is not great since the good ones go abroad soon after their PhD degrees. What does one do to have a research group at my age?

In May 2014, a new government headed by Prime Minister Narendra Modi came

Prime Minister Narendra Modi

into power. Modi sent for me soon after he took over and talked to me for thirty minutes. Modi has announced several major national missions and priorities. One is awaiting new programmes and initiatives in science and technology devoted to the development of the country. I am now free of any major involvement in science and technology policy and planning.

In November 2014, the first Commonwealth Science Conference was held in Bangalore under the auspices of the Royal Society. I was chairman of the organizing committee along with Tony Cheetham. The conference was a great success academically and culturally. I gave one of the plenary lectures at the inaugural session, after the lecture by Sir Paul Nurse, President of the society. Based on the success of this meeting, it has been decided to have biannual commonwealth science conferences as a programme of the Royal Society.

Whatever happens in future, I want to continue doing research. The real concern, however, is whether my research output will be affected because of the shortage of students and other factors. I could increase science outreach activities to keep busy, but that cannot compensate for my interest in chemical research. I hope and pray that it will be possible for me to continue to be as active scientifically as I have been till now.

In science outreach activities, Indu has been a companion and an active partner. I am delighted that her untiring efforts were recently recognized by the Karnataka Women's University, which awarded her an honorary DLitt degree. The National Academy of Science, Allahabad, has elected her as a fellow in recognition of her service to science and society.

Age reminds me of a funny incident in Winston Churchill's life. When Churchill turned eighty, his portrait was painted by a young artist. After he completed the painting, the young artist thanked Churchill and said, 'Sir Winston, thank you for the opportunity to paint your portrait. I hope that I will get to paint your portrait again when you are ninety.' Churchill looked at the young artist and said, 'I don't see why not! You look quite healthy.'

Harry Kroto with a model of C$_{60}$

Even as I worry about the possible limitations that I may face due to my age, I must record that I am receiving an increasing number of invitations to deliver lectures in symposia and conferences all over the world. As an illustration, let me list what I had to do between June and December 2015. In June 2015, I had to go to London to give a talk at a Royal Society meeting on catalysis. In July, I gave talks at a symposium in Cambridge, followed by one in London. The London symposium was organized to celebrate the thirtieth anniversary of the discovery of fullerenes by Kroto and others. It was sad to see Harry Kroto physically disabled due to motor neuron disease. Harry was younger than me and a friend for a long time. He has since passed away. In September, I gave the Wigner memorial lecture in

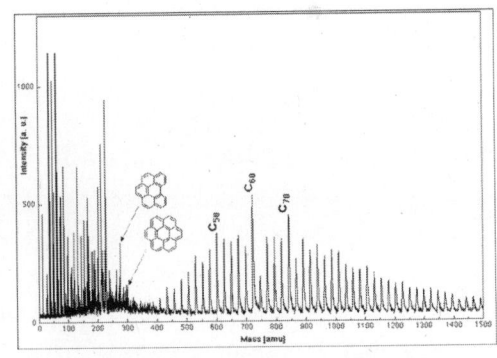

Mass spectrum of the vapour of carbon

the US and was also the distinguished visiting lecturer at Temple University. In November 2015, I gave a talk at the Sorbonne, and two weeks later, had to be in Vienna for the TWAS meeting.

In December, I went to Canberra to be admitted to the Australian Academy of Sciences as a corresponding member and receive an honorary doctorate from the Australian National University. I enjoyed giving a talk at the university. In January 2016, I had to go to Taiwan to talk at a symposium and receive an honorary doctorate degree. These foreign visits become taxing since they do not minimize one's commitments within the country. I have decided that after 2016, I will be very careful in committing myself to foreign visits as well as to lectures within the country.

In closing

I can characterize my life after sixty as follows. It has been wonderful and most enjoyable. While the flow of research papers has certainly been a factor leading to this happy state, the major contributor to all that I have been able to do is my wonderful wife, Indu. She is positive in her attitude and enjoys life as one ought to. We have similar interests in reading and music. We start each day by listening to music and Sanskrit chants early in the morning, along with a cup of coffee. Music has been a great source of joy and peace. I listen to music at lunch time, when I close my office door for about forty minutes. Music and science have certain similarities; both are limitless with regard to excellence. Other than music, I love cooking as a hobby. I am a reasonable chef and can cook a variety of dishes. The beauty of cooking is that one knows the result right away by looking at the faces of those who eat. I also read fiction, historical novels and biographies as much as I can. With cooking, music, visiting the Himalayas, searching for tigers and a wonderful family, my science has flourished after my sixtieth year. What else would one want in life?

Rao in the kitchen as Tony Cheetham watches

'As one goes on doing unfettered scientific research, science becomes part of life; it becomes a way of life'—Anon

'There is only one solution to India's problems and that is science, more science and still more science'—C. V. Raman

7. MY HEROES

One's life is enriched by learning about the lives of great men and women. They influence the way one functions and operates in this world. In this respect, I have been influenced by some of the greatest personalities of India, thanks to my mother who talked about them incessantly when I was younger. She talked of the great Buddha, who, 2500 years ago, found the cause of human suffering. She talked of Adi Shankaracharya, the most cited author in India, who propounded the highly original Advaita philosophy 1300 years ago. Adi Shankara went to Badrinath and other places in the Himalayas by foot, all the way from Kerala, twice in his short lifetime. There have been many other great philosophers that India has produced. Madhvacharya, for instance, gave importance to the real world where we live, in addition to the world of *adhyatma* (spirituality). Then, Swami Vivekananda in the twentieth century gave a new dimension to our understanding of India. I have adored Mahatma Gandhi, whom I saw once when I was younger. It still amazes me that he galvanized an entire subcontinent without the aid of telephones and aeroplanes. Gandhiji showed how tenacity, dedication and doggedness are essential if one has to achieve a major goal. One requires these qualities for success in science as well.

I have found in the last few years that most children get excited about science when they see a nice experiment or listen to a brilliant lecturer. This was true in my own case, when I heard Professor C.V. Raman at the age of eleven. In science, there is continuity of knowledge and tradition. We can never forget that we stand on the shoulders of great scientists who made major discoveries in the past. In this context, I must record my admiration for the great contributions of some of our Indian scientists.

J.C. Bose discovered telegraphy in the late nineteenth century in Presidency College, Kolkata. Is it not amazing that this Bengali physicist made such a major discovery in a college in Kolkata even before the discovery of the electron? He demonstrated this to the public by doing the experiment in the open. He wrote a paper on the subject in 1895, whereas the electron was discovered only

J.C. Bose

in 1897. Unfortunately, he was not awarded the Nobel Prize and Marconi was given all the credit. J.C. Bose taught physics to two young students, S.N. Bose and M.N. Saha, at Presidency College, both of whom went on to become famous physicists. Working in India, S.N. Bose made the path-breaking discovery of Bose–Einstein statistics, by collaborating with Einstein.

How did C.V. Raman accomplish so much, considering he only worked on his scientific research after office hours? Much of his early scientific contributions were made when he went to a college laboratory after finishing his days' work as an officer in the accounts department of the British government. At Nagpur Science College, he made major discoveries in acoustics while working on musical instruments. This earned him the fellowship of the Royal Society. He spent his evenings working in the Indian Association for the Cultivation of Science, in Kolkata, and the work he did here earned him the Nobel Prize.

How did Ramanujan, living in a small place in south India, come up with so many mathematical theorems and conjectures, which made him one of the greatest mathematicians of all time? He had no research grants or PhD students. In fact, he did not go to college, and yet, he became an FRS. What Robert Kanigel wrote about Ramanujam describes the situation perfectly.

'In 1913, a twenty-five-year-old Indian clerk with no formal education wrote a letter to G. H. Hardy, then widely acknowledged as the premier English mathematician of his time. Srinivasa Ramanujan begged Hardy's opinion regarding several ideas he had about numbers. Hardy realized that the letter was the work of a genius. Thus, began one of the most productive and unusual scientific collaborations in history, that of an English don and an impoverished Hindu genius

Srinivasa Ramanujan

whose like has never been seen again. Hardy arranged for Ramanujan to sail for Cambridge, leaving behind his wife and mother in Madras. Ramanujan's isolation from his family and the intensity of his work eventually took their toll, and within seven years of leaving India he was dead. For Hardy, "the collaboration with Ramanujan was the one truly romantic incident of my life."'

In my professional life, I have admired some fine scientists for their contributions and dedication to science. Professors Nevill Mott and H.C. Brown have been my role models. I have also adored many chemists from the early years, starting with Lavoisier (1743–94), who is considered the father of chemical science. Here, I present brief sketches of three of my heroes in chemistry—Faraday, Lewis and Pauling.

Michael Faraday (1791–1867)

One of my heroes, ever since I was a child, has been Michael Faraday. He is considered one of the greatest scientists of all time. He was born on 22 September 1791 in England, to a poor family. After three years of schooling, at the age of fourteen, he had to take up a job to support his family. He worked as an apprentice to a bookbinder, where he started reading books on science and developed an interest in the subject. One of the books he read was Marcet's *Conversations in Chemistry*. Accidentally, he got an opportunity to listen to three lectures on chemistry by Sir Humphry Davy in 1812. Luckily, he managed to get a job as an attender in Davy's laboratory at the Royal Institution—he was required to clean glassware and assist Davy in his experiments. He went with Davy on his European tour as his assistant. Slowly, Faraday developed skills to conduct experiments on his own. His first research paper was published in 1816, when he was twenty-five years old. He was fascinated by experimental work and started exploring a variety of problems covering topics of interest in chemistry, physics and engineering.

(From Wikimedia Commons)

As a chemist, Faraday made many discoveries—the most famous one being the laws of electrolysis (1833–36). He discovered benzene (1823) and also worked on optical glass, catalysis and liquefaction of gases. He was the first one to make nanoparticles of metals in 1857. In physics, Faraday's contributions are enormous. He discovered electricity by electromagnetic induction (1831). He studied phenomena like electromagnetic rotation and demonstrated the importance of static electricity. He showed by a simple experiment that oxygen was paramagnetic. His experiment on Faraday rotation is truly ingenuous. Most importantly, he was the first one to talk about the unification of forces through his experiments on the lines of force. Einstein held him in such regard that he had Faraday's photograph in his office and felt that he was the father of the idea of unification of forces. Faraday's ideas were provided physical and mathematical basis by Maxwell. A person with only three years of schooling became a fellow of the Royal Society in 1825. When he was thirty-four years old, he received an honorary doctorate from Oxford University, and is the only scientist to receive the Royal Medal and the Copley Medal of the Royal Society twice.

Faraday was a simple man with great ideas. His simplicity extended to his personal life. He refused the knighthood from the Queen and the Presidentship of the Royal Society. He was well-known for answering difficult questions simply. When the Chancellor of the Exchequer asked Faraday what the practical use of his discovery of electricity was, he said, 'But, sir, there is every probability that you will soon be able to tax it!' When the Prime Minister asked him the same question, he asked, 'What good is a newborn baby?' Faraday regularly gave lectures to children and the public. His lecture on the chemical history of a candle is well known. Faraday was a great communicator and his advice to speakers was not to give lectures that last for more than an hour.

Faraday's work ethic was unbelievable. He worked very hard every single day. For example, during 1833, he carried out experiments on 24 December. The next entry in his record book was on 26 December, with 25 December missing as it was Christmas. He published over 400 research papers single-handedly, each with a new idea or finding. Lord Rutherford said that had Faraday lived in the twentieth century, he would have probably received six Nobel Prizes.

Faraday passed away on 25 August, 1867 at the age of seventy-six.

'The world of science lost on Sunday one of its most assiduous and enthusiastic members. The life of Michael Faraday has been spent from early manhood in the single pursuit of scientific discovery and though his years extended to 73, he preserved to the end the freshness and vivacity of youth in the exposition of his favourite subjects, coupled with a measure of simplicity which youth never attains . . . as a man of science he was gifted with the rarest of felicity of experimenting, so that the illustrations of . . . subjects seemed to answer with magical ease to his call. . . . He was one of those men who have become distinguished in spite of every disadvantage of origin and of early education, and if the contrast between the circumstances of his birth and of his later worldly distinction be not so dazzling as is sometimes seen in other walks of life, it is also true that his career was free from the vulgar ambition and uneasy strife after place and power which not commonly detract from the glory of the highest honours. No man was ever more entirely unselfish, or more entirely beloved. Modest, truthful, candid, he had the true spirit of a philosopher and of a Christian, for it may be said of him, in the words of the father of English poetry,

"Gladly would he learn, and gladly teach."

The cause of science would meet with fewer enemies, its discoveries would command a more ready assent, were all its votaries imbued with the humility of Michael Faraday'—*The Times*

'There was a philosopher less on earth, and saint more in heaven' — J.H. Gladstone

G.N. Lewis (1875–1946)

G.N. Lewis is one of the greatest chemists of all time. He described chemistry as a subject which 'encompasses everything that is interesting'. He is one of the great pioneers who brought research and education in chemistry to the forefront. He was born in 1875 in the US and received his BSc and PhD degrees from Harvard University. Unfortunately, the professor with whom he worked at Harvard was not generous and some of his ideas were not duly supported. For example, shortly after the discovery of the

(From www.pbs.org)

electron, Lewis wrote a paper on electrons and the molecule, but it did not receive fair recognition. If it had, the idea of the chemical bond may have emerged earlier than 1916.

After some years at MIT, Lewis went to the University of California, Berkeley, in 1912 to establish a new chemistry department. The department he built continues to be the greatest chemistry department in the world. He created an atmosphere where research is of paramount importance, while promoting the teaching of modern chemistry. The most important contributions of Lewis were to chemical thermodynamics. He made the subject part of chemistry. He worked on free energies of chemical substances and on verification of the third law. He corrected the third law of thermodynamics proposed by Nernst. In 1916, he wrote the classic paper on the chemical bond (the centenary of which was celebrated in 2016), recognizing the role of valence electrons. His idea of the chemical bond included the octet theory. He wrote the first books on chemical thermodynamics and valence (the structure of atoms and molecules). Another major contribution of his was the description of new types of acids and bases—Lewis acids and Lewis bases—based on whether molecules donate or accept electrons. He did considerable work on various aspects of heavy water. He initiated work on photochemistry and on certain aspects of electronic spectroscopy.

Lewis changed his research fields effortlessly, and his contributions to physical chemistry and chemical physics were ground-breaking. The name 'photon' was suggested by him. He was one of the founders of the *Journal of Chemical Physics*. Linus Pauling dedicated his famous book *The Nature of the Chemical Bond* to Lewis.

Pauling wrote to Lewis on 29 August 1939: 'I am very happy to know that you are pleased with my book, and feel that it is good enough to be worthy of its dedication to you. You know, of course, that I had you in mind continually while it was being written, and I have been hoping that my treatment would prove to be acceptable to you.'

Lewis had written to Pauling earlier on 25 August 1939, 'I have just returned from a short vacation for which the only books I took were half a dozen detective stories and your *Chemical Bond*. I found yours the most exciting of the lot.'

For reasons that nobody understands, Lewis was not awarded a Nobel Prize even though he was nominated nearly thirty times. His contributions to thermodynamics were ignored. The idea of the chemical bond was not considered important enough. Lewis acids and bases did not impress the Nobel committee, nor did his contributions to heavy water and photochemistry. His students Seaborg, Urey and Calvin, all received Nobel prizes.

Lewis was an extraordinary person in many ways. He worked hard every day and would dictate papers to students after dinner at night, drawing on the data from memory. He rarely travelled. To compensate for his shyness and his limited public appearances, he had a fantastic style of written English that one enjoys even today. Just to illustrate his style of writing, I give below a few sentences from the preface of his classic book on thermodynamics:

> There are ancient cathedrals which, apart from their consecrated purpose, inspire solemnity and awe. Even the curious visitor speaks of serious things, with hushed voice, and as each whisper reverberates through the vaulted nave, the returning echo seems to bear a message of mystery. The labour of generations of architects and artisans has been forgotten, the scaffolding erected for their toil has long since been removed, their mistakes have been erased, or have become hidden by the dust of centuries. Seeing only the perfection of the completed whole, we are impressed as by some superhuman agency. But sometimes we enter such an edifice that is still partly under construction; then the sound of hammers, the reek of tobacco, the trivial jests bandied from workman to workman, enable us to realize that these great structures are but the result of giving to ordinary human effort a direction and a purpose.
>
> Science has its cathedrals, built by the efforts of a few architects and of many workers.

G.N. Lewis was a born experimentalist. He died in 1946 after performing an experiment. He had asked his PhD student, Michael Kasha, to record a spectrum of a compound in liquid HCN. Since Kasha was not enthusiastic, Lewis offered to prepare liquid HCN for him. When Kasha returned to the laboratory after lunch, he found an ampoule of HCN in the vacuum line and a broken ampoule on a table. The laboratory smelled of HCN. Lewis was found dead sitting in a chair. Was it suicide or an accident? One does not know. As Pitzer said, Lewis was such a brilliant man that he could make suicide look like an accident.

'Gilbert Newton Lewis typified the physical chemist of great intuition who was able to conceive of beautifully simple models and concepts to explain complex physical and physic-chemical phenomena . . . His own career touched every aspect of science and in each he left his mark . . . He epitomized the scientist of unlimited imagination and the joy of working with him was to experience life of the mind unhindered by pedestrian concerns . . .' (from Kasha).

'As a man, Lewis was a great soul whose inspiration will never be forgotten by those who knew and loved him. He was one of those rare scientists . . . who are also great teachers and leaders of a school, so that their influence is multiplied by the many they have inspired . . .' (from Gettell et al).

Linus Pauling (1901–1994)

As has often been said, there was one Linus Pauling and there will be no other. Pauling was born on 28 February 1901 in the US. His parents were immigrants and moved from one place to another. His father owned a drugstore and was very good to young Pauling. He died at a young age, when Pauling was only nine years old. Pauling had to start earning money by taking up odd jobs. He even laid roads to earn wages to help

(From http://lpi.oregonstate.edu)

his mother. In spite of these difficulties, he went on to study chemical engineering and obtained a BSc degree from Oregon State University. He met his future wife at university and she was a great support to him in all his endeavours.

During his undergraduate studies, Pauling read the famous papers of G.N. Lewis and Langmuir on chemical bonding. He entered Caltech for his PhD degree, where the head of the department (A.A. Noyes) took an interest in him and advised him to take up research on the determination of structures of crystals with X-rays. This paved the way for Pauling interest in structural chemistry. He solved many structures even as a PhD student. After obtaining the PhD degree, encouraged by Prof. Noyes, he went to Europe for postdoctoral work with Arnold Sommerfeld. He learnt quantum mechanics and wrote a seminal paper on the theoretical prediction of physical properties of many electron-atoms and ions in 1927. This paper was not liked by the old guard, which felt threatened by him.

Linus Pauling returned to Caltech as a member of the faculty in the chemistry department and continued his phenomenal work in structural chemistry. He solved a variety of structures using X-ray diffraction and electron diffraction, and came out with a number of concepts in structural chemistry. It is at this time he derived the Pauling rules and defined electronegativity. He elaborated on the structure of the chemical bond, starting with the early inputs of G.N. Lewis and propounded the valence bond theory. He authored the famous book *The Nature of Chemical Bond* in 1939. He also wrote the first book on quantum mechanics. While his work was on molecules and minerals to start with, he made a foray into biology and explained the cause of sickle cell anaemia—a molecular disease. In 1951, he solved the structure of the alpha helix. This was a major contribution and marks the beginning of molecular biology. When Pauling was asked how he came up with such a great idea, he said 'to have a great idea, one needs to have many good ideas'. He was a person with many ideas.

Pauling received the Nobel Prize for chemistry in 1954, for his work on the nature of the chemical bond. Around this time, he got involved in societal problems, in relation to nuclear testing in the atmosphere. Pauling played a major role in the campaign against atmospheric nuclear testing and this invited the wrath of the US government. His passport was withdrawn as the government considered him a security risk. His stand on nuclear testing earned Pauling the Nobel Peace Prize, but this was not endorsed by the institutions and the US government. The Nobel Prize for peace to Pauling was considered by many as an insult to the United States. Pauling went through a very difficult time when he had to leave Caltech and the American Chemical Society, and move from one place to another. By the mid 1970s, things changed and he was accepted by society and fellow professionals again.

In 1952, there was an important meeting on DNA in London. Pauling could not attend this meeting since he had been denied a passport. At that meeting, X-ray photographs of DNA taken by Rosalind Franklin were shown. If Pauling had gone to that meeting and seen those pictures, it is likely that he would have solved the structure of DNA as well (and got the third Nobel Prize).

Pauling's contributions to chemistry are not only through research but also through chemical education. He was an exciting communicator. He wrote the first textbook on general chemistry for beginners. Pauling's life can be characterized in the following way: he was an extraordinary scientist with great ideas and a fertile imagination, and a humanist with high values. He was a fearless crusader on issues that he believed in.

Pauling's advice to young people was, 'When an old and distinguished person speaks to you, listen to him carefully and with respect—but do not believe him. Never put your trust in anything but your own intellect. Your elder, no matter whether he has grey hair or has lost his hair, no matter whether he is a Nobel laureate or not, may be wrong. So you must always be sceptical— always think for yourself.'

The last few years of Pauling's life were not happy ones, particularly after his wife died. She was a champion of peace and gave great support to Pauling in his endeavours. Pauling referred to her as his auxiliary memory bank.

'I have been especially fortunate for about 50 years in having two memory banks available—whenever I can't remember something I ask my wife, and thus I am able to draw from this auxiliary memory bank. Moreover, there is a second way in which I get ideas. I listen carefully to what my wife says, and in this way, I often get a good idea. I recommend to young people that you make a permanent acquisition of an auxiliary memory bank that you can become familiar with and draw upon throughout your lives.' (from Goertzel and Goertzel).

Pauling passed away in 1994 at the age of ninety-three.

'The world lost one of its greatest scientists and humanitarians, and a much respected and beloved defender of civil liberties and health issues. Because of his dynamic personality and his many accomplishments in widely diverse fields, it is hard to define Linus Pauling adequately. A remarkable man who insistently addressed certain crucial human problems while pursuing an amazing array of scientific interests, Pauling was almost as well known to the American public as he was to the world's scientific community.' (from Marinacci, *Linus Pauling – In Memoriam*)

In closing

Chemistry is a vast subject with a large number of outstanding contributors. Chemistry made a beginning as a subject only in the eighteenth century after Lavoisier. Chemistry was dominated by structure and bonding till about the 1960s, but in recent years, the scope of the subject has gone beyond the molecular frontier. A number of new areas—such as self-organization, synthetic biology, complex systems, atmosphere, advanced materials, and so on—have emerged. Computation has contributed in a big way to chemistry, while experimental work has become very sophisticated thanks to present-day instrumentation. Although this is not the place to write about the great scope of chemistry, I am mentioning this to show how a subject changes over the years. In the early days, when I started working in chemistry, it was based on simple experiments mainly using test tubes and beakers, and with little or no instrumentation. Today, demands on the quality of experiments are increasing day by day. This is true of physics and other subjects as well. It is difficult to predict the future directions in science, and their possible impact on society and on science itself. There will certainly be great contributions from many heroes. Those future heroes are children today playing somewhere in the world, possibly in Asia (in India!).

'In the temple of science, there are many mansions, and various indeed are they that dwell therein and the motives that have let them thither. Many take to science out of a joyful sense of superior intellectual power; science is their own special sport to which they look for vivid experience and the satisfaction of ambition; many others are to be found in the temple who have offered the products of their brains on this alter for purely utilitarian purposes. Were an angel of the Lord to come and drive all the people belonging to these two categories out of the temple, the assemblage would be seriously depleted, but there would still be some men of both present and past times left inside'—Albert Einstein

EPILOGUE

PERSONAL REFLECTIONS

I am now in my eighty-third year of existence. I did not think I would be writing a book about science at this age, but I continue to enjoy the life of a scientist. It has been a wonderful life. Someone asked me a while ago what I would do if I were to do it all over again. I would love to be a PhD student in the US, and start publishing again. I have no regrets on any account and am grateful for all the opportunities I have had to work and serve.

Whatever I have tried to do in science, has been possible because of the extraordinary support that I have received directly or indirectly from several sources. India, which I love, has done everything possible to keep me afloat and prosper professionally. I have enjoyed contributing to science planning when I have been asked to do so by the government of the day, but I am equally happy just working in a laboratory with my students. It has been a pleasure to have been involved in establishing two important professional societies, the Chemical Research Society and the Materials Research Society. Both are thriving.

I have often been asked whether I ever guessed that I would accomplish as much as I have. Of course I had no inkling of how I would perform in life. I certainly did not expect to receive seventy honorary doctorate degrees or the membership of most major science academies in the world. Neither did I expect to keep publishing continuously for over sixty years. It is my earnest feeling that the support of my family and the contributions of my students are responsible for enabling me to get to this stage in my professional life.

I have had a large number of PhD students and post-doctoral fellows. Without exception, they have all worked hard and have been very loyal. I have always stayed in close touch with my co-workers on a day-to-day basis and have written papers with them. When a research problem is at a crucial stage or is yielding interesting results, I see my students every day. Staying in contact with my students on an everyday basis has minimized the wastage of time and has increased research productivity. One of the probable reasons why I have been able to publish many papers is because of this relationship with students. This does not mean that I dictate everything that happens in the laboratory. We discuss my ideas as well as those of the students and come to a decision on what exactly we have to do in the laboratory. Working with students has helped me stay young. Most of my students have stayed on in academic professions and some of them have become famous. Several of them have been elected to science academies and have become directors of laboratories. Some of them have received major recognitions in terms of medals and prizes.

I have tried to encourage my students by giving them first authorships and I also try helping them with their personal problems when required. I have probably inherited this quality from the fine teachers I have had. All the professors with whom I have worked, particularly in the US, were very kind to me and helped me to develop as a scientist. They allowed me to publish papers independently. They were generous with their time, and I have tried to emulate these qualities of generosity and kindness.

I constantly remember my fine parents, who never imposed their will on me and allowed me to make up my mind on all matters, particularly with regard to my academic future when I was young. Not once did my parents suggest I take up an administrative position or something else, but took pride in the fact that I was a scientist and a professor. I must also remember my teachers who gave me all their love and encouragement. I express my gratefulness to all those who have helped me and extended support in my professional endeavours. I am thankful that many academies and professional organizations have found me worthy of various recognitions. I do realize that the rocks of accomplishment will be washed away by the waves of time (as Tagore wrote), and yet the awards bring cheer and happiness.

My wife has been my rock, never once discouraging me from what I have been doing. She has been my auxiliary memory bank— always cheerful and positive. She has been an active participant in programmes related to science education. She has not bothered me about banking, shopping and various household matters. Professor H. C. Brown often said, 'I only do two things to keep happy. One is to publish research papers and the other is to keep my wife happy.' I have followed this advice.

With Suguna and Simba

With Indu and Sanjay

We have wonderful children and grandchildren, and we often holiday together. Our daughter, Suchitra, is busy taking care of the family and she works in a nursery school. Her husband, Dr K.N. Ganesh, heads the Indian Institute of Science Education and Research in Pune. Our son, Sanjay, works in multimedia. Our grandson, Kartik, is a doctor and is going to be a nephrologist. He grew up in our house and we have enjoyed his company. Our granddaughter, Suguna, is always in touch with us via mobile phone and is studying economics. Then, there is our dog, Simba, a wonderful member of the family. Our great-granddaughter has just started making noises to give advance notice of the future.

I have worked with several scientists from different countries in my long career. In this process, some scientists from Europe, America and elsewhere have become very close friends. Some friends like Tony Cheetham and Mike Klein have

With Professor and Mrs H.C. Brown

**With Tony
Cheetham in Egypt**

**With Mike
Klein**

become part of our family. Science does breed good friendship and
bonhomie. One has to be open-minded to enjoy these wonderful
benefits of science. International collaboration and working with
people from other countries has also benefited me and helped
me broaden my horizons in science. Working in international
organizations like IUPAC, TWAS and ICSU has helped me
identify myself with the world community of scientists. One
becomes a world citizen in this process and is at home anywhere
in the world.

Science is changing with time. What used to be considered
chemistry at one time is no longer the chemistry of today.
Chemistry today is highly multidisciplinary or cross-disciplinary.
If one wants to do really good research today, one must know
about several other disciplines. Interesting chemistry is often

done by those who work with physics, biology, materials and so on. Strong links between theory and experiment is another aspect which characterizes modern physical science. Many of my papers have contributions from theoretician colleagues and this collaboration has improved the quality of my work, giving more meaning to the experimental results.

The volume of science being published today has increased beyond all expectation. The super-exponential increase in publications is staggering. Not only has it made it impossible to keep up with the literature, but also extremely difficult to decide on the problems to work on. The explosive production of literature must be disheartening to young scientists, especially those from developing countries. The rush to publish has also made refereeing difficult for journals. When I started my career, I tried to pick an area of research where I could contribute significantly. I looked for an area which was not crowded and was guided by Robert Frost's famous poem.

> Two roads diverged in a yellow wood
> And sorry I could not travel both
> I took the one less travelled by
> And that has made all the difference.

Picking a lonely road is still a good way to choose a research area, but this is becoming increasingly difficult nowadays. I have been lucky that solid state and materials chemistry was in its infancy when I started working more than half a century ago. It is a mainstream subject today, with a large number of practitioners. I am now treated as a grandfather of the subject. I must narrate a funny incident in this connection. A few years ago, I was in a European airport with a friend, after giving a plenary lecture at an international conference. A gentleman came rushing to me and said, 'You may know a person with a similar name as yours, who had done work on oxides and

related materials about thirty or forty years ago. He must be dead by now.' I had to tell him that I was the person who wrote those papers.

Scientific research has become extremely competitive in recent years—a problem acutely felt by us in India, especially because of the high performance by our Asian neighbours, China, Japan and South Korea. I remember the China of the early 1980s, when the conditions there were worse than those in India. China has made massive investments and is publishing almost the same number

of research papers as the US today. I am not too bothered by the number of papers India produces, but I worry about the quality of these papers. We have to improve the quality substantially to be able to compete in research.

In this scientific scenario, one finds it more and more difficult to publish papers in good journals. Journals are categorized in terms of the impact index and if one wants to publish in a high-impact journal, life is made more difficult. Some of the journals use more referees than one would want. There are journals which use four to six referees, possibly to find some reason or the other to reject the paper. There are some journals which return papers to authors even without refereeing them. The sheer number of manuscripts received appears to be beyond the capacity of journals to treat them appropriately. This has created a difficult situation especially for young scientists, who may get desperate because of the way their papers are treated. Continually receiving negative reports on manuscripts tires people. I have found a simple solution for myself. I do the best possible work and send it to a good subject journal and not worry about publishing in the so-called famous journals with popular acclaim. As Professor Mott often used to tell me, if the work is good enough, people will read the paper. I have not done badly for having published in professional journals.

While academic research remains an essential and significant part of scientific effort, there are many other demands on science as well. We have to work on pressing problems of humankind such as energy, environment, water, health, and food security. Industries need crucial scientific inputs. Science is also the starting point of much innovation and start-ups are generally based on cutting-edge research from a laboratory. We need to take care of all these aspects of scientific effort for progress.

How does one do science successfully in India? How does a young person ever reach high levels in science in today's competitive world? The problems are severe and the solutions require contributions from many sources. First of all, our institutions have to be improved substantially in terms of infrastructure and facilities. It is distressing that even today I cannot depend on the city's electric supply for my research. Reliable electric supply has been a problem since I started my research career in 1959. Regarding research facilities, I had to wait for nearly twenty years to have reasonable support. There should be more institutions of excellence in India to accommodate the increasing number of talented people. Our universities, which have decayed over the years, need to be rejuvenated on a war footing. Funding for education, which has stagnated around 2–3 per cent of GDP of the central budget, needs to be increased. We have to increase it to 6 per cent of the GDP. Of the 2 per cent of the GDP provided for education, less than 0.5 per cent is probably used for higher education. How can this ever support IITs and universities? How can an IIT become an MIT?

We have to seriously concern ourselves with educational and employment opportunities for the large body of young students. How will we take care of the additional few million who will enter higher education in the next decade or so? We need to plan carefully. An important aspect relates to the search for talent. More than 60 per cent of young Indians are in rural India and there must be another Ramanujan or Raman there. We have to tap this talent. I consider this to be as important as any other national endeavour. Talent search has to receive greater importance for India to shine. Innovation is eventually done by individuals and we need the most talented people for this purpose. When my wife and I travel to various parts of interior India, from Uttarakhand to Kerala, we see so many young people excited about science and education. We have to provide them the right opportunities.

Funding for science and technology has been poor. It has never been more than 1 per cent of the GDP. The central government has to increase it to 1.5 per cent or 2 per cent of the GDP as early as possible. The industry has done little for scientific research and has to contribute to it significantly. Only around 10 per cent of science in India is supported by the industry. Instead, it should be closer to 40 or 50 per cent of the scientific effort as in other countries.

A problem that we have faced since Independence relates to the undue importance given to big science, while small science is ignored. Whenever one mentions science, one mentions atomic energy, space, and such agencies. However, real progress in science occurs through small science. It is small science that improves the quality of life and solves problems related to health, energy and other areas, crucial for human survival and prosperity.

Bureaucracy has been, by and large, a negative factor affecting science. For example, the rules and procedures employed for scientific institutions are the same as those for a district office. Scientists are forced to retire at the age of sixty or so. There is no age limit for scientists in the US and many other countries. It is extremely important that good scientists are able to contribute to national efforts as long as they are able to do good work. Procedures for monitoring projects are so archaic that people do not receive funds regularly. Projects get closed before the last year's funding is provided because of the need of some certification or the other.

I have till now talked only about the external factors related to scientific performance. However, there are other, equally important, factors that affect performance. In my experience, almost all the great scientists that I have known worked very hard. Prof. Mott in Cambridge or Prof. H.C. Brown in Purdue (both Nobel laureates), and many other fine scientists I have known, worked till the last day of their lives. They were both past ninety when they died, and even published papers during the last year of their lives. G.N. Lewis and Michael Faraday worked extremely hard to accomplish major results in science. We have to work much harder. Chinese scientists seem to work much harder than us. More importantly, we should not forget that there are many things that we can do without money.

I have noticed some change in the attitudes of young people in the last few years. Unlike in the early years, some of the younger

people today are far too ambitious and anxious. Some of them
want instantaneous fame. Desire for administrative positions is
an evil that seems to poison young minds. Some of them want to
become rich at a young age. Nowadays, among youngsters, I see
a tendency to travel, often for unimportant purposes. Selfishness
in scientists, young or old, is harmful for the growth of science.
However selfish young people may be, senior scientists have to
be generous and make sure that they provide all the help and
encouragement so the future of science is safe.

In any intellectual effort, selfishness interferes with one's
performance in a big way. One can be fearless only if one is not
selfish. I would like to quote a poem from Tagore's *Gitanjali* in
this connection.

> I came out alone on my way to my tryst
> But who is this that follows me in the silent dark?
> I move aside to avoid his presence
> but I escape him not.
> He makes the dust rise from the earth
> with his swagger; he adds his loud voice
> to every word that I utter
> He is my own little self, my lord, he knows
> no shame; but I am ashamed
> to come to thy door in his company.

Another factor that interferes with quality work is our obsession
with trivial things. Some people worry about the price of onions
more than their research. Trivia interfere with higher things in
life. Here, again, there is a wonderful poem from the *Gitanjali*
that I would like to quote.

> This is my prayer to thee, my lord – strike, strike at
> the root of penury in my heart
> Give me the strength lightly to bear my joys and sorrows

Give me the strength to make my love fruitful in service
Give me the strength never to disown the poor or
bend my knees before insolent might
Give me the strength to raise my mind
high above daily trifles.
And give me the strength to surrender my
strength to thy will with love

Anger and envy are inimical to creative work. One cannot be angry all the time and still do wonderful work in the laboratory. I have joked about this matter often. One cannot fight with one's wife and do science that day in the laboratory. These points may look unimportant, but are not to be ignored. Envy and jealousy dominate the actions of many individuals. We should be generous and ensure that others thrive along with us. It is particularly true of senior people, who should never become envious of bright young people. I am reminded of Swami Vivekananda's statement that envy among people may not allow India to become great.

A quality that is essential for the progress of Indian science is tolerance for failure. Failure is a part of success. Without failure, one cannot find great success in science and it is important that we learn not to look down on failure. I have noticed that some people wait for good people to fail or make mistakes. This appears to be a quality common to all sections of society (including politicians). One seems to enjoy talking about the failures and misfortunes of others. In science, we cannot afford this luxury.

I have said little about our society itself. Indian society as a whole has to start viewing science as an important aspect of national effort. Science has to find a place in our value system. Suicides, rape and murder, cricket, banking, interest rates and foreign investment, all occupy prime places in the media. Society seems obsessed with matters related to money and business. Science, environment, education and scholarship should become important aspects of our value system and receive due attention. If scientists, and the society, pay heed to some of the points discussed here, nothing can stop India from becoming a scientifically advanced country. At that time, there will be more institutions in India that are comparable to MIT or Cambridge and will be among the top fifty institutions in the world.

In closing

With all the problems and challenges that I have mentioned above, I cannot say that life has been dull. I have actually enjoyed growing up with science in India, and building laboratories and institutions. Professional loneliness, which was a serious limitation in the early years of my career, has diminished. Although the main river of science flows from Europe to America, I have been able to find a place in the sun. In this respect, I owe much to my students for their contribution, a large body of scientist friends abroad and a loving family.

As a proud Indian, I feel good that all my research was carried out in India. I have managed to find a balance between my research efforts and my contributions to institution building and science in the country. I have sincerely served many scientific organizations and agencies such as atomic energy, CSIR, the Planning Commission and the University Grants Commission. I have served IITs not only as a professor but also as the chairman of the Standing Committee of the IIT Council for an extended period. I have also worked with institutions such as TIFR. I have not, however, sought an administrative position. I did not take up the several offers to become a member of the Rajya Sabha. Research has been my sole passion in a subject I have loved. Talking to young children in their formative years through science outreach programmes has added another dimension to my life. Contact with young people and with scientists around the world has helped me be a citizen of this complex world, concerned about the future. I cannot think of a better way to live.

At a personal level, I have been a happy person and have no complaints. I have enjoyed knowing people like Satish Dhawan, who was the moral conscience of the scientific community. I have had the pleasure of working with or knowing many fine people, some of whom have become close friends. I have not disliked anyone intensely, and have done my best to be helpful and useful to people. I am truly sorry if I have hurt anyone unintentionally. I have not uttered a lie in my life (as far as I can remember) and my wife vouches for this assertion. I have not attended family functions and other social occasions, and have not visited relatives. I am truly sorry for this omission and can only make up for this lapse in my next life. This is a promise!

I cannot forget certain events in my professional life. I remember travelling to Mumbai from IIT Kanpur for discussions with Vikram Sarabhai on issues of mutual interest, which could not happen since he had passed away the previous night in Trivandrum. I waited at the airport to see him for

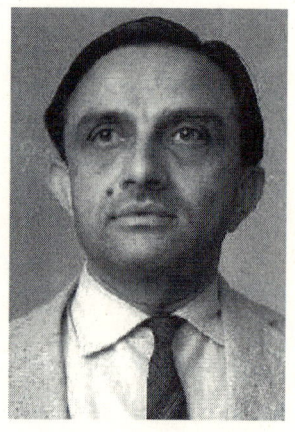

Vikram Sarabhai

the last time. When Homi Bhabha visited IIT Kanpur, he asked me to meet him

Homi J. Bhabha

in Mumbai after he returned from a trip abroad. Unfortunately, his plane crashed in the Alps on that trip.

A great Indian scientist I think of frequently is G.N. Ramachandran. I feel that GNR was not treated well by the scientific community and others. The only consolation I have is that I could get him back to IISc when I was director by creating a unit for him. He was a simple person and was somewhat child-like. GNR was the first one to inform me of my impending election to the Indian Academy of Sciences in 1964 and the first one to congratulate me when I was elected to the Royal Society. I was terribly upset on seeing him in his last days.

I do not worry much about myself. The only worries that bother me are related to my research or science in the country. It is common for me to compose sentences or plan contents of an article through the night. A matter from the past that I keep remembering is from the last meeting of the Indian Academy of Sciences that Professor C.V. Raman attended. During the tea-break, Professor Raman told me 'Professor Rao, I am eighty-one years old. It bothers me that

Solid state chemists in Tuscany (2002)—John Goodenough is fourth from the left. Tony Cheetham is next to me. To his right is Raveau

India is not on the top of the world in science.' I am now eighty-two years old and I echo Professor Raman's statement.

As I get older, I miss my old friends. Almost all my teachers and some of my friends from my student days in the US and elsewhere are all gone. I miss dear friends like Satish Dhawan and Raja Ramanna. I cherish the memories of my professional friends of the yesteryears. Paul Hagenmuller and John Goodenough are over ninety-four years

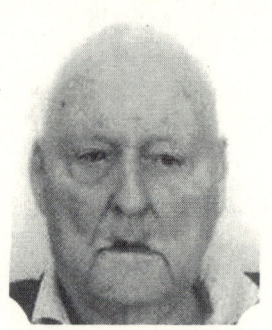

Paul Hagenmuller

old, and Paul has been an invalid for some years now. The only solid state chemist from the old days with whom I am in contact with is George Honig at Purdue. M.G.K. Menon is ill and is in no position to converse. S. Varadarajan too is not in a good state. There are only few old friends left.

Whatever be the problems, one has to have a sense of humour to survive in this world. I find it enjoyable to recollect statements of great men in this regard. One such is the following statement of Einstein:

Two things are infinite;
the universe and human stupidity.
I am not sure of the universe.

It is time to say goodbye. I wish everyone a great future and the aspiring scientist, great science. Following one of Tagore's statements, I feel that science has given me much and that I have given little to science. When asked what he asks of god, the great music maestro Bismillah Khan said, 'Oh lord, keep me in the world of music till the end.' My prayer is, 'Oh lord, keep me in the world of science till the end.'

I end with a song from the *Gitanjali*:

Where the mind is without fear and the head is
held high
where the knowledge is free
where the world has not been broken up into
fragments
by narrow domestic walls;
where words come out from the depth of truth; where
timeless striving stretches its arms towards perfection;
into the dreary desert sand of dead habit;
where the mind is led forward by thee into
ever-widening
thought and action
Into that heaven of freedom, my Father,
Let my country awake

—Rabindranath Tagore

APPENDIX 1

LETTER TO A YOUNG FRIEND

My dear young friend,

I have received your letter expressing your fears and aspirations. Your letter clearly shows that you are concerned about your studies and also about the possible opportunities for young people in science and other areas. I understand your fears. I do not blame you because the situation is confusing. In many universities, where admissions are based on the marks you get, the system has gone haywire. I learn from newspapers that there are students not getting admissions to colleges even after obtaining 99 per cent marks in examinations. There are also too many competitive exams for admissions to national institutions such as IITs and IISERs, as well as other institutions. A student interested in admissions to a medical college ends up taking innumerable national- and state-level competitive exams. Admission to IITs seems to be essentially determined by coaching institutes. I do hope that we will soon have a system wherein intelligent boys and girls can easily pass the entrance exams of IITs and other institutions because of their merit. This was the situation till a few years ago. I am terribly upset that our entire education system has become examination oriented. 'Education' has lost where 'examinations' have gained.

You wanted to know about the choice of courses and the difficulty in deciding on a course of study. Let me tell you about the areas of interest today. If you are really interested in science, take up science. This would be ideal if you like the challenge of discovery and innovation. Science provides immense opportunities for one's creative contributions. Today's science is different and is highly multidisciplinary. It is only those who really understand several areas of science that contribute to major advances.

For example, chemistry today is no longer the chemistry of the old days. Real contribution to chemistry today is possible only when you also have an understanding of biology, physics and so on. There are people who say that great contributions in biology come from those who know a lot of chemistry. Materials science is an interdisciplinary subject, which makes use of chemistry, physics, biology, engineering and so on. I, therefore, leave it to you which science subject you would like to pick as your speciality, but please make sure you have a good foundation with regard to different sciences and other core subjects, which could include computer science and some topics in engineering science. If you are interested in science, try to get into IISER, IISc or one of the IITs, which have a science undergraduate programme. There are also some colleges and universities with good science degree courses. If you cannot get into an integrated master's degree programme, do not worry. Get a bachelor's degree and then go for an integrated PhD programme (which gives a master's degree after the first two or three years and then a PhD). Otherwise, get an MSc from a university and try to go to the place of your choice for PhD work. In any case, during your undergraduate studies try for a summer research fellowship offered by the science academies and this centre (JNCASR). Remember that there are many challenges in science and you can use your talent to solve problems in the frontiers of science, and also those related to the pressing problems of mankind.

If you are interested in engineering, you could first try to get into the IITs, NIT or elsewhere, but do not worry too much if you do not get admitted to one of the institutions that you would like to join. Go to a reasonably good institute and get a good basic degree in engineering. Even after engineering, you can change over to something else. If you do not like to be an ordinary engineer, but do scientific research, it is certainly possible. Some of the greatest scientists had basic degrees in engineering. Linus Pauling got his bachelor's degree in chemical engineering and became one of twentieth century's greatest scientists with two Nobel Prizes. The same is true of Robert Mulliken, another Nobel laureate, who obtained an engineering degree from MIT and later received a Nobel Prize in chemistry. I am assuming that you would be interested in doing research from what you say. Anyone with a sound background in science or engineering can pursue research even at a later stage. Research requires a certain attitude and not years of preparation. If one is excited about learning or discovering something, there will always be openings somewhere. All one has to do is to pick up some knowledge of the subject before embarking on research.

If you become a medical doctor, and in case you want to be a research-oriented doctor, do not get bogged down by clinical practice. Let me tell you the story of a friend of mine. He got a BA degree in English from Harvard University. After getting a BA, he felt that it was not enough to be an English teacher. He obtained a master's degree in English. After that he felt what he was interested in was biology and medicine. So, he took some courses in biology and got admitted into a medical college. He got an MD degree and later worked in some hospitals, including a period he spent in India. He then started doing research on cancer. The research he did on cancer earned him a Nobel Prize. This is the life story of Harold Warmus. I wish children in this country get opportunities of this kind, where they can readily switch from one subject to another. What we badly require is a flexible curriculum where young students can take a package of courses of interest, which will allow them to switch from one area to another easily. This will also help them move from one institution to another. The highly rigid undergraduate curricula that we have today in most institutions do not allow such flexibility.

It is best not to worry too much. I am sure that you will find a place in the sun. Keep working hard and try your best. Something will work out. Do not knock yourself out trying to get admissions. Do not get disheartened and depressed if you do not get admission to the institution that you want for undergraduate studies. Do well in whatever you take up and remember that the main qualities required for success are determination, dedication and tenacity.

I wish you success in whatever you decide to do. Please do not hesitate to write to me if you have any doubts or queries. Once again, do not forget that any one determined to accomplish something in life will do so.

With best wishes.

Sincerely yours,
C.N.R. Rao

APPENDIX 2

MANY A GREAT SCIENTIST GAVE THEIR ALL TO SCIENCE AND TO THE YOUNG

Alexander von Humboldt was born in Berlin in 1769. Natural science was yet to develop at that time. When Humboldt died in 1859, experimental research was in vogue—to a large extent due to Humboldt himself. He laid the foundation of many areas of science, including those in modern geology, geography, space science and the study of nature. Humboldt supported a number of scientists including Justus von Liebig, Joseph Louis Gay-Lussac and Carl Friedrich Gauss, clearly due to his farsightedness. There is a story of an American visitor named Taylor, who went to see Humboldt, who was then very old. As Taylor was leaving, Humboldt said, 'You have travelled much and seen many ruins. Now you have seen one more' and gave his hand. Taylor replied, 'Not a ruin, but a pyramid', for he pressed the hand which had touched those of Frederick the great, of Forster, of Klopstock and Schiller, of Pitt, Napoleon, Josephine, Jefferson, Wieland, Herder, Goethe, Cuvier, Laplace, Gay-Lussac—in short, of every great person of that period.

Alexander von Humboldt

(From Wikimedia Commons)

(From Wilhelm Ostwald, Wikipedia)

Wilhelm Ostwald

Ostwald was a great teacher of chemistry and is considered to be the founding father of physical chemistry. What is interesting is that in the history of physical chemistry, nobody has trained a larger number of great physical chemists than Ostwald. Almost all the early physical chemists of the United States worked with him. These include people like Theodore William Richards, G.N. Lewis, Richard M. Noyes and so on. Ostwald was an extraordinarily kind person. When Svante Arrhenius was without a place to go after his doctorate degree and was desperate, Ostwald accommodated him in his laboratory. Walther Nernst also worked there at that time. Ostwald received the Nobel Prize in 1909.

Lord Rutherford did great research even as a young man and received the Nobel Prize for chemistry as early as 1908. He then came out with the structure of the atom in 1911. He established the Cavendish Laboratory in Cambridge as the greatest centre of physics of that period. What is interesting is that he encouraged so many persons to do work in his laboratory, many of whom ended up receiving Nobel Prizes. Some of the greatest names in physics such as John

Cockcroft, Ernest Walton, James Chadwick, Patrick Blackett and Pyotr Kapitza were all his students. Niels Bohr considered Rutherford as his teacher. Rutherford took great pride in the fine work done by his students and encouraged many young people to do research in the Cavendish. There was a joke about the laboratory at that time. Taxi drivers would not take any one to go there. They would say we don't like to go near that building, because if we go there, Rutherford will make us to do some research and we will have to win the Nobel Prize.

Ernest Rutherford

(From Wikimedia Commons)

G.N. Lewis is one of the greatest chemists who did not get the Nobel Prize, but many of his students received the award and a large number of great chemists were produced in his laboratory. The Nobel Prize-winning students of Lewis include Glenn T. Seaborg, Melvin Calvin and Harold Urey. Arnold Sommerfeld in physics was similar. Some of the greatest physicists worked with him as doctoral students. These include luminaries such as Werner Heisenberg, Wolfgang Pauli, Hans Bethe and Pauling. Sommerfeld, the great teacher of physics, was ignored by the Nobel committee.

Arnold Sommerfeld

(From http://timerime.com)

The future has always belonged to the youth.
To the young, I say:

'I . . . express a wish that you may in your generation, be fit to compare yourself to a candle; that you may, like it, shine as lights to those about you; that in all your actions, you may just by the beauty of the taper by making your deeds honourable and effectual in the discharge of your duty to your fellow men'—
Michael Faraday

Be yourself.
   You are a child of the universe, no less than the trees and the stars; you have a right to be here. With all its sham, drudgery and broken dreams, it is still a beautiful world.
   Be careful. Strive to be happy.

From desirerata found in
St. Paul's Church, Baltimore, 1692

To strive for knowledge is to

struggle for freedom; for, ignorance

is the worst slavery

—C.V. Raman

We are scientists, but

we are first human beings

—Anon

SOURCES

'Albert Einstein – Biographical'. *Nobelprize.org.* Nobel Media AB 2014. Web. 28 Apr 2016. http://www.nobelprize.org/nobel_prizes/physics/laureates/1921/einstein-bio.htm

'Paul J. Crutzen – Facts'. *Nobelprize.org.* Nobel Media AB 2014. Web. 18 Apr 2016. http://www.nobelprize.org/nobel_prizes/chemistry/laureates/1995/crutzen-facts.html

'Vitaly L. Ginzburg – Facts'. *Nobelprize.org.* Nobel Media AB 2014. Web. 18 Apr 2016. http://www.nobelprize.org/nobel_prizes/physics/laureates/2003/ginzburg-facts.html

'Walter Kohn – Facts'. *Nobelprize.org.* Nobel Media AB 2014. Web. 18 Apr 2016. http://www.nobelprize.org/nobel_prizes/chemistry/laureates/1998/kohn-facts.html

'Jean-Marie Lehn – Facts'. *Nobelprize.org.* Nobel Media AB 2014. Web. 18 Apr 2016. http://www.nobelprize.org/nobel_prizes/chemistry/laureates/1987/lehn-facts.html

'Abdus Salam – Facts'. *Nobelprize.org.* Nobel Media AB 2014. Web. 17 Apr 2016. http://www.nobelprize.org/nobel_prizes/physics/laureates/1979/salam-facts.html

http://www.sci.ccny.cuny.edu/~sarachik/ (From Sarachik Homepage)

'Richard P. Feynman – Biographical'. *Nobelprize.org.* Nobel Media AB 2014. Web. 28 Apr 2016. http://www.nobelprize.org/nobel_prizes/physics/laureates/1965/feynman-bio.html

File: Ghhardy@72.jpg, From Wikimedia Commons, (source: A mathematician's apology, 1927)

File: Sir CV Raman.jpg, From Wikimedia Commons, 1930, source: From Nobel Lectures, Physics 1922-1941, Elsevier Publishing Company, Amsterdam, 1965.

www.penguinbooksindia.com

From www-outreach.phy.cam.ac.uk

Jnan Chandra Ghosh DSc, F.N.I. (From Wikimedia Commons)
'Herbert C. Brown – Facts'. *Nobelprize.org.* Nobel Media AB
2014. Web. 16 May 2016. http://www.nobelprize.org/nobel_
prizes/chemistry/laureates/1979/brown-facts.html
Kenneth Pitzer, source: http://www.lbl.gov/images/PID/Pitzer.
html
G.N. Lewis, From scarc.library.oregonstate.edu (Source: Edgar
Fahs Smith Collection, Schoenberg Center for Electronic Text
and Image, University of Pennsylvania Library)
Oxford University — www.meetuniversity.com & www.
explorewithed.co.uk_
Sir Nevill Francis Mott (1905-1996), British Physics and Nobel
Laureate (1977). From Wikipedia, the free encyclopedia
G_NRamachandran.jpg ,From Wikipedia, the free Encyclopedia,
Picture of Gopalasamudram Narayana Iyer Ramachandran,
Current Science, Vol. 80, no. 8, 25 April 2001, pg 908.
From Wikimedia Commons By Dmitry Tonkonog – Own
work, CC BY-SA 3.0, https://commons.wikimedia.org/w/index.
php?curid=27973817
jthomas.jpg (From www.clarkson.edu)
Cshekhar & IK Gujral — http://pmindia.gov.in
C.V. Raman picture: www.younews.in
Pictures of UCSB: www.ucsb.edu & oep.ucsb.edu
Picture of Narendra Modi: twitter.com
Harry Kroto at Sussex University with a model of the C60
molecule Credit: Connors Brighton published in *The Telegraph*,
Sir Harold Kroto Obituary, 2 May 2016.
Acharya Jagadish Chandra Bose Source: http://www.vigyanprasar.
gov.in (Author: Vigyan Prasar, Government of India)
Picture of Srinivasa Ramanuja: www.usna.edu
picture of Michael Faraday, 1861, Source: Opposite p. 290 of
Millikan and Gale's *Practical Physics* (1922) From Wikimedia
Commons.

G.N. Lewis: www.pbs.org

Linus Pauling: http://lpi.oregonstate.edu

Vsarabhai .jpg — www.thehindu.com

Bhabha.jpg — From Wikimedia Commons

R-Tagore — From Wikimedia Commons

Alexander von Humboldt, Artist: Stieler, Joseph Karl (Alexander von Humboldt – 1843.jpg, From Wikimedia Commons).

Wilhelm Ostwald, From 'Wilhelm Ostwald', Wikipedia

Ernest Rutherford LOC.jpg, New Zealand chemist and Nobel Prize laureate Ernest Rutherford (1871-1937) Author: George Grantham Bain Collection (Library of Congress), (From Wikimedia Commons)

Arnold Sommerfeld — From http://timerime.com

ABOUT THE ACADEMY

The Indian Academy of Sciences is a scholarly society founded by the Nobel laureate Professor C.V. Raman in 1934. The academy aims to promote the practice of all areas of sciences in pure as well as applied aspects.

Since its founding, the academy has worked to promote original research and the dissemination of scientific knowledge through a variety of activities that include publication of scholarly journals, organizing scientific meetings and seminars, and, in recently years, has undertaken an annual effort in science education.

One of the major concerns of the academy is to promote science education nationwide among bright students and motivated teachers throughout the country to attain a certain level of excellence. Each year, about 2000 Summer Research Fellowships are offered competitively to science students and teachers to enable them to undertake research internships in laboratories in universities and research institutes across the country. The academy also organizes numerous refresher courses and lecture workshops that are aimed towards the improvement of science education and teaching in the country. Details of the various activities of the academy can be found on the website http://www.ias.ac.in

Professor C.N.R. Rao was President of the Indian Academy of Sciences between 1989 and 1991, and was instrumental in initiating many of the science education as well as science outreach activities of the academy.